WRITING
TO LEARN
BIOLOGY

RANDY MOORE

Wright State University

Universidad Católica, Santiago, Chile

W R I T I N G

T O L E A R N

B I O L O G Y

Saunders College Publishing

A Harcourt Brace Jovanovich College Publisher

Fort Worth   Philadelphia   San Diego   New York   Orlando   Austin   San Antonio

Toronto   Montreal   London   Sydney   Tokyo

Text Typeface: ITC Garamond
Compositor: Digitype, Inc.
Acquisitions Editor: Julie Alexander
Managing Editor: Carol Field
Project Editor: Anne Gibby
Copy Editor: Becca Gruliow
Manager of Art and Design: Carol Bleistine
Art Assistant: Caroline McGowan
Text Designer: Tracy Baldwin
Cover Designer: Lawrence R. Didona
Text Artwork: Rolin Graphics
Director of EDP: Tim Frelick
Production Manager: Charlene Squibb
Marketing Manager: Marjorie Waldron

Printed in the United States of America

WRITING TO LEARN BIOLOGY

ISBN: 0-03-074189-0

Library of Congress Catalog Card Number: 91-050654

2345 039 987654321

PREFACE

It is with words that we do our reasoning, and
writing is the expression of our thinking.
  —W.I.B. Beveridge, *The Art of Scientific
    Investigation*

A writer is a person for whom writing is more
difficult than it is for other people.
  —Thomas Mann

Learn as much by writing as by reading.
  —Lord Acton

Language is the only instrument of science, and
words are but the signs of ideas.
  —Samuel Johnson

This is a book about writing to learn biology. I wish it could be short and
simple like some "scientific writing" books, but I wanted to provide more
than a pocket-sized list of writing clichés such as "Be precise" and "Write
concisely." Such simplistic reminders may help experienced writers, but

mean little to novices. Indeed, telling an inexperienced writer to "Write concisely" is like telling a baseball player to "Hit the ball squarely." They *know* that—what they *don't* know is *how* to do it. Memorizing simplistic rules will not ensure good writing. Writing well requires that you understand the *process* of writing. Only then will you understand why some writing communicates clearly while other writing does not, how word choice affects the impact of a sentence, and why much of "scholarly" writing is clutter that should be deleted to enhance communication. Understanding the process of writing goes beyond obeying a set of grammatical commandments, and will show you that good writing is clear thinking on paper. This is why any understanding necessary to produce good writing must be based on a coherent system of principles more useful than lists of anecdotes, platitudes, clichés, and self-evident truisms such as "Be specific."

But why another book about scientific writing? Most books about scientific and technical writing assume that the sole purpose of writing is to communicate. Consequently, these books seldom go beyond describing how to write a research paper and how to avoid problems of punctuation, grammar, and style. They include all of the clichés about accuracy, brevity, clarity, and cohesion, but typically provide only examples rather than an understanding of effective writing. Such one-sided accounts of writing fail to square with expert writers' advice on how to write effectively. Moreover, although these books gesture toward their audience, few help you learn to anticipate what your readers want and are looking for.

Knowing something about grammar and how to write about biology is important for communicating biological information. Consequently, these topics are included in *Writing to Learn Biology*, although in a unique format. However, I contend that communication is not the *primary* goal of writing. The primary goal of writing, like reading, is to help you learn and understand. Writing is as much an instrument of thinking as it is a silent language that we use to pursue our thoughts. Moreover, only after you understand a subject can you make that understanding available to others in writing. Consequently, good writing should communicate *and generate* ideas. As James Van Allen said, "I am never as clear about any matter as when I have just finished writing about it."

Most books about scientific writing fail to explore writing as a way of thinking and learning. This limited approach to writing ignores *composition*, the essence of thinking. Although one function of writing is communication —the clear and direct expression of our ideas—many of our thoughts exist only when put to paper. By failing to link writing and thinking, we fall into a frustrating trap called "writer's block." This and other problems such as the obsession to "get it all on paper at one time" result from not understanding the writing process, and underlie the embarrassment and self-delusion of comments such as "Well, that's not what I really meant" and "I understand the topic but just can't find the right words."

Writing can be fun and exciting. Millions of people do it for therapy and entertainment. However, many students (and professors) find writing tedious and difficult because they take a "think now and write later" approach to writing. This usually results from dwelling on the product or mental origins of writing, to the exclusion of the *activity* of writing. This approach handicaps you because it separates writing from thinking, thereby robbing you of writing's most valuable gift: its ability to help you learn. Writing is a powerful way of thinking that can help you understand a subject, define a research problem, and hone an experiment. However, thinking an assignment completely through before writing about it pressures you to get it all done at once—to get all of the ideas onto paper in an organized way at one sitting. Such an approach is virtually impossible.[1] A much more enjoyable and effective approach is to *view writing as thinking*—to write as you think things through. When approached this way, writing becomes a tool for communicating and for learning, because writing one sentence and paragraph suggests other sentences and paragraphs. This approach turns writing into discovery and helps you think with your pen, piece of chalk, or keyboard.

Learning About Writing

The inability of most students to enjoy writing or to write effectively attests to the failure of most university writing programs. These programs fail because they usually concentrate only on having students write without regard to helping them understand writing. Students already know how to write— they've cranked out all sorts of essays, book reports, and 20-page term papers for years. This kind of exercise hasn't improved students' writing because it has not taught them to write *effectively*. Most writing assignments have been little more than unpleasant exercises in practicing poor writing.

Although students often have an uneasy sense about what they write, they seldom know how to identify and correct the problems in their writing. Consequently, they become like drivers who know that the car won't start but have no idea about what to do to get it started. Although these people dutifully raise the car's hood, their hit-and-miss fiddling with wires and battery connections seldom starts the car and often causes even more problems. Similarly, students who rely only on simplistic rules and clichés rather than an understanding of writing usually just scramble words and sentences in hopes of improving their writing. At best, this produces frustrated students who wonder why their grades do not improve despite their having followed all of the "rules." This emphasizes an important point: Even when

---

[1]To worry about saying something and about saying it well are too much to worry about at one time. The results of trying to do all this are predictable and mimic those of one of Albert Camus' characters in *The Plague*. This character begins writing without considering his audience or doing any other planning. Throughout the novel, Camus describes the character's frustrating attempts to write perfectly by fiddling with word choice and syntax. The character's novel never gets beyond the first sentence.

you don't feel anything is wrong with your writing, others often do. To prevent this problem, learn how to be the first critic of your writing—how to anticipate a reader's difficulties and to hear yourself as others hear you. You'll do this not when you memorize rules, but when you understand writing and readers' expectations.

Writing can be fun and educational. However, like other tools of a professional, writing is difficult to master because it requires clear, sustained thinking. Even great writers complain that writing is difficult. For example, Nathaniel Hawthorne wrote in his journal, "When we see how little we can express, it is a wonder that any man ever takes up a pen a second time." Similarly, Ernest Hemingway said, "What you ultimately remember about anything you've written is how difficult it was to write," and Flannery O'Connor wrote, "All writing is painful and if it is not painful, then it is not worth doing." Great biologists, too, have realized that good writing takes much time and is hard work, as evidenced by this quote by Charles Darwin: "A naturalist's life would be a happy one if he had only to observe and never to write." If you think that writing is often hard work, you are in good company.

Writing is often difficult because it, like science, requires that you simultaneously be imaginative and critical. The imagination and discovery inherent in writing about biology involve exploration, risk-taking, and leaps into the unknown. Conversely, the self-criticism necessary to revise a paper or experiment requires detachment, skepticism, and testing. This tug-of-war between discovery and criticism represents the True Believer and Doubting Thomas in each of us. Since these two mighty antagonists are not easily appeased at the same time, writing about biology often seems difficult. However, so are many things that we learn to do well and come to enjoy, such as riding a bicycle, skiing, and playing a musical instrument. So take heart—writing is less a matter of mystery than a mastery of skills, many of which you already have. I hope that *Writing to Learn Biology* will help take the sting out of learning to write well by helping you understand the process of writing.

Learning to write effectively doesn't require a life-long, monklike dedication. Rather, all it requires is that you want to write well and that you understand the process of writing. Understanding this process will help transform your writing from a helter-skelter "get it all on paper" chore to something you do to learn, to communicate, and even to relax.

What to Expect in This Book

I wrote *Writing to Learn Biology* to show students, biologists, and other professionals how writing about biology improves learning about and communicating biology. You'll see that the essence of effective writing is based first on discovering your ideas, second on understanding the many decisions that writers make, and third on mastering the skills that translate those decisions into writing that communicates effectively. Contrary to what you may think, none of these decisions is hard to make, nor are the steps that turn these decisions into a well-written product. The trick is knowing what decisions to make, where those decisions will take you, and mastering the appropriate techniques. When you understand the process of writing, you'll trust the process because it will tell you what you know, what you don't know, and what you need to know. Good writers trust this process because they know it will eventually lead them to a clear and focused paper. Conversely, the inability to write effectively often typifies inexperienced writers who don't understand the process of writing and don't know where the process will lead.

Understanding writing will make writing easier than you ever thought it could be. It will also have other benefits, such as helping you write better term papers, lab reports, and answers on "essay" exams. However, give yourself time to learn and understand the process. Waiting until the last minute to start writing makes this process, and good writing, out of the question. At best, you'll end up with a paper that a critic might describe as "extremely difficult, but only somewhat rewarding for the persistent and dedicated reader." Since most scientists are neither persistent nor dedicated readers, you'll probably get a poor grade on the paper or exam. Similarly, a poorly written research paper will probably be rejected or ignored.

I wrote *Writing to Learn Biology* not as a writing "expert," but rather as a biologist who views writing as a useful and necessary tool of an effective scientist. I have tried to write a readable and accessible "how to" manual that is educational and practical. I hope that *Writing to Learn Biology* will help you view writing as a process rather than as a product. Moreover, I've assumed that you want to write better in a short time and without much fuss. Consequently, I've emphasized the simple and practical aspects of writing, and have tried to organize those ideas in a guide that leads you through the process of writing, shows you the purposes that writing serves, helps you improve your writing, and shows you how to use writing as a tool for learning. I stress what's immediately useful to most writers—the minimum amount of understanding needed for self-defense against common errors that impede learning and communication. This involves understanding not how sentences and paragraphs work within grammatical theories, but how to write effectively about biology in the real world. I've left the abstract rules and explanations to the several excellent guides to grammar available at most bookstores (see Appendix 1).

The guide listed below summarizes effective writing and will help acquaint you with this book. It is also a good place to quickly evaluate your writing skills. If you're a novice writer, you'll probably benefit most by reading the chapters in the order in which they're presented. If you're an experienced writer, use this guide to shape the first and later drafts of your paper, to rethink your ideas, and to clarify the goals and importance of your work:

Content

Subject
Have you defined your subject? See page 36.
Have you gathered your ideas? See page 31.

Audience
Have you identified your audience? See page 37.
Does your writing match that audience? See page 37.
For nonscientists, have you defined all terms and provided examples? See page 207.
For experts, have you presented enough information? See page 234.
Have you answered questions that your readers will likely ask? See page 50.
Are your conclusions supported by evidence and logic? See page 50.

Bias
Have you used inappropriate stereotypes or labels? See page 179.

Logic
Is your writing logical and coherent? See page 165.
Did you have a plan? See page 48.
Did you follow your plan? See page 50.

Interest
Did you emphasize important conclusions? See page 172.
Did you vary the lengths of your sentences? See page 172.
Did you write effective paragraphs? See page 166.
Did you make the reader want to read on? See page 173.

Precision and Clarity
Did you use simple words and sentences? See page 69.
Did you provide details to support generalizations? See page 170.
Did you use specific rather than vague words to ensure precision? See page 114.
Will readers understand all of the pronouns you've used? See page 126.
Have you punctuated your writing to help readers grasp your ideas? See page 181.
Have you used familiar words? See page 88.
Have you organized your thoughts logically? See page 48.

Grammar

Have you removed dangling modifiers? See page 123.
Are modifiers close to the words they modify? See page 124.
Have you expressed similar ideas in similar ways? See page 129.

Style

Usage

Have you used the right word? See page 114.
Have you avoided clichés? See page 126.
Have you used active voice? See page 137.
Have you avoided doublespeak? See page 116.
Have you avoided excessive hedging? See page 176.

Conciseness

Have you eliminated all unnecessary words? See page 75.
Have you eliminated all of the "fuzz" and filler from your paper? See page 69.

Readers' Expectations

Have you followed the subject as soon as possible with its verb? See page 132.
Have you used a strong verb to express the action of every clause and sentence? See page 135.
Have you put in the stress position the material that you want the reader to emphasize? See page 140.
Have you put familiar information at the beginning of the sentence? See page 142.

Mechanics

Have you defined all of the abbreviations that you've used? See page 128.
Have you checked the spelling of all words? See page 260.
Have you used numbers correctly? See page 286.
Are all of the tables and figures necessary? See page 294.
If so, are they numbered and well designed? See page 294.
Have you asked a colleague to read your paper? See page 185.

For students: Many of you may dread using this book, primarily because you equate writing with a long list of rules called "grammar." Your experiences with writing were probably not too pleasant, nor was sitting through those seemingly endless lectures on grammar and "rules of writing." Many of your English teachers probably taught you how to search for hidden meanings in poetry or to read literature instead of how to write effectively. Moreover, you've probably been told to strive for a "scholarly" style of writing rather than a strong and unpretentious style that will carry the thoughts of the world you live in. Consequently, by now you may be somewhat cynical and distrustful of any kind of writing or "writing book." I went through the same kind of training—uninspiring classes and countless 20-page term papers in which the most important criteria were the paper's length and format rather than its message. Therefore, I don't blame you for being cynical about the goals of this book. Considering your experiences, it probably has not occurred to you that writing can be a fun and useful, if not critical, tool that could make or break your career.

I hope that *Writing to Learn Biology* will change your attitude by showing you that writing can be exciting, enjoyable, and easily understood. I also hope that this book will help you overcome the often overwhelming and largely unnecessary emphasis on "dos and don'ts" associated with good writing. This book does not teach writing as a set of rigid rules—such an approach bores most people and has little, if any, lasting effect. Indeed, if all it took to write well was a list of rules, all scientists would write well. This is certainly not the case. Moreover, obeying all of the rules included in many writing books can produce wretched prose—that's why to many of the most lucid and precise writers, many "rules" have no standing whatsoever. Rather than concentrate on rules, *Writing to Learn Biology* stresses practical writing —the kind you need to discover your ideas, to get them onto paper, and to efficiently convert those ideas into concise, well-written sentences and paragraphs. To do this, you must be able to diagnose problems and correct them, not just list rules.

Once you understand how to write well, you'll like writing for the same reasons you like biology. They're both exciting, creative, lively, and inspiring; they are full of action and verifiable, defined facts; they require precision and argumentation. Moreover, learning to write well will make you a better biologist because it will show you how biologists think and work. Specifically, it will show you the nature of biology, help you design better experiments, help you write better term papers, and move you away from blind acceptance to real data, interpretation, and argumentation.

This book does not pretend to teach creative writing, and I know of no magic formula to convert everyone into a Hemingway. However, this book will teach you useful, effective writing—the kind of writing that will produce papers that are clear, forceful, precise, well organized, and easy to read and understand. *Writing to Learn Biology* includes exercises to help you write more effectively. It also stresses the importance of revision—preparing several drafts of a paper—to effective writing and learning. This approach to writing will reduce your anxiety about writing and show you that effective writing can be learned by understanding and applying a few principles of writing.

In *Writing to Learn Biology* you'll read many of the classic papers and essays about biology. These papers have diverse styles and purposes—for example, Rachel Carson's writing tries to persuade you to act, Stephen Jay Gould's writing describes riddles of evolution, and James Watson's writing describes the thrill of scientific research. All of these papers will make you think. Moreover, all will show you how writing can help you learn about biology. You'll also read some bloopers written by students, faculty, and others. For example:

After soaking in acid, I washed the glassware thoroughly.

Life begins at contraception.

Madman Curie discovered radio.

Samuel Morse invented a code of telepathy.

Gravity was invented by Isaac Walton. It is chiefly noticeable in the autumn, when the apples are falling off the trees.

Magellan circumcised the globe with a hundred foot clipper.

I hope these bloopers will teach you about writing. I also hope they make you laugh.

Whether this is your first or last course in biology, *Writing to Learn Biology* will help you better understand biology and other sciences. It is meant as a handbook, so keep it handy. Write as you use this book—use it as a writing coach and reference to guide you through your writing projects. Keep a marking pen handy to underline or highlight points that you especially want to remember. Moreover, don't wait until you've read the whole book to start practicing what you learn. Remember the saying "What I read, I forget. What I see, I remember. What I do, I understand."

I hope you enjoy the book. Let me hear from you.

Randy Moore
Dayton, Ohio
October 1991

ACKNOWLEDGMENTS

According to one bit of wisdom,

Good things, when short, are twice as good.
   —Baltasar Gracián

So I'll be brief.

With *Writing to Learn Biology* I've tried to give teachers, students, and others a usable and entertaining book about how to use writing to learn biology. Some material in this book is new, some is old, and the rest is borrowed. I've cited sources where I could, and apologize for those parts whose origins I can't recall. If they're especially good, I've assumed that they were my ideas.

This book required much work and I am grateful to many people for their help. I thank Cathleen Petree for signing and encouraging me to write this book, and Julie Alexander, my editor, for patiently guiding me through this project. I greatly appreciate the freedom she gave me as I wrote this book. Darrell Vodopich, Jim Seago, Judy Verbeke, Joyce Corban, Dennis Clark, and Kris Moore reviewed the manuscript; their comments helped shape the book in important ways. Joyce Corban tolerated my daily rantings about writing, biology, and life. Every writer should have such a helpful friend and colleague. I also thank my students who have taught me much about biology and writing.

Most of all, I thank my family and friends for their support and encouragement. This book is for Mom, Dad, and Kris.

R.M.

CONTENTS

UNIT ONE

WRITING
ABOUT
BIOLOGY

. . . [W]riting comes in grades of quality in the
fashion of beer and baseball games: good, better,
best. . . . Better ways can be mastered by writers
who are serious about their writing. There is
nothing arcane or mysterious about the crafting of
a good sentence.
　　—James Kilpatrick

Young writers often suppose that style is a garnish
for the meat of prose, a sauce by which a dull dish
is made palatable. Style has no such separate
entity; it is nondetachable, unfilterable. The
beginner should approach style warily, realizing
that it is himself he is approaching, no other; and
he should begin by resolutely turning away from all
devices that are popularly believed to indicate
style — all mannerisms, tricks, adornments. The
approach to style is by way of plainness, simplicity,
orderliness, sincerity.
　　—E.B. White

Reading maketh a full man, conference a ready man,
and writing an exact man.
   —Francis Bacon

The scientific man is the only person who has
anything new to say and who does not know how
to say it.
   —Sir James M. Barrie

Your knowing is nothing, unless others know you
know.
   —Persius

If thought corrupts language, language can also
corrupt thought.
   —George Orwell

Everything that can be thought at all can be
thought clearly. Everything that can be said can be
said clearly.
   —Ludwig Wittgenstein

CHAPTER ONE

What is

Effective Writing?

It is impossible to dissociate language from science
or science from language, because every natural
science always involves three things: the sequence
of phenomena on which the science is based; the
abstract concepts which call these phenomena to
mind; and the words in which the concepts are
expressed. To call forth a concept a word is needed.
   —Antoine Lavoisier

The more students write, the more active they
become in creating their own education: writing
frequently . . . helps students discover, rehearse,
express and defend their own ideas.
   —Toby Fulwiler

Put it before them briefly so they will read it,
clearly so they will appreciate it, picturesquely so
they will remember it and, above all, accurately so
they will be guided by its light.
   —Joseph Pulitzer

Biology is a process that helps us learn about our world. Biologists design experiments to test hypotheses and use the results of experiments to learn about life. In doing so, biologists face a dual challenge—to understand what they're doing and to communicate this understanding to others. Biologists must create meaningful descriptions of their ideas so that readers can easily re-create, and therefore understand, the ideas. These challenges mean that a biologist must know how to do and write about science.

Many biologists equate effective writing with "correct" writing—the kind of writing that breaks none of the magical rules they memorized in their English classes. To see the fallacy of this logic, read the following essays without pausing too much, and then consider your impressions of the quality of each writer as a scientist:

Brown's Version

In the first experiment of the series using mice it was discovered that total removal of the adrenal glands effects reduction of aggressiveness and that aggressiveness in adrenalectomized mice is restorable to the level of intact mice by treatment with corticosterone. These results point to the indispensability of the adrenals for the full expression of aggression. Nevertheless, since adrenalectomy is followed by an increase in the release of adrenocorticotrophic hormone (ACTH), and since ACTH has been reported (*Brain*, 1972) to decrease the aggressiveness of intact mice, it is possible that the effects of adrenalectomy on aggressiveness are a function of the concurrent increased levels of ACTH. However, high levels of ACTH, in addition to causing increases in glucocorticoids (which possibly account for the depression of aggression in intact mice by ACTH), also result in decreased androgen levels. In view of the fact that animals with low androgen levels are characterized by decreased aggressiveness the possibility exists that adrenalectomy, rather than affecting aggression directly, has the effect of reducing aggressiveness by producing an ACTH-mediated condition of decreased androgen levels.

Smith's Version

The first experiment in our series with mice showed that the total removal of the adrenal glands reduces aggressiveness. Moreover, when treated with corticosterone, mice that had their adrenals taken out became as aggressive as intact animals again. These findings suggest that the adrenals are necessary for animals to show full aggressiveness.

But removal of the adrenals raises the levels of adrenocorticotrophic hormone (ACTH), and *Brain* (1972) found that ACTH lowers the aggressiveness of intact mice. Thus the reduction of aggressiveness after this operation might be due to the higher levels of ACTH which accompany it.

However, high levels of ACTH have two effects. First, the levels of glucocorticoids rise, which might account for *Brain*'s results. Second, the levels of androgen fall. Since animals with lower levels of androgen are less aggressive, it is possible that removal of the adrenals reduces aggressiveness only indirectly: by raising the levels of ACTH it causes androgen levels to drop.[1]

Both essays present the same information in the same order and use the same technical words. Both essays are also "correct"; they differ only in their use of ordinary language. However, Smith's essay is more readable because it avoids unfamiliar words, avoids inflated roundabout phrases, and uses shorter, more direct sentences. Conversely, Brown's essay is hard to read because it contains long sentences, big words, and convoluted constructions. These differences have a tremendous impact on other scientists. Indeed, almost 70 percent of the 1,580 scientists who read these essays judged Smith's essay as more stimulating, more interesting, more impressive, and more credible than Brown's essay. Readers also thought Smith was more helpful, dynamic, and intelligent than was Brown. Most important, when asked to judge Smith's and Brown's competence —specifically, which biologist seemed to have a better-organized mind— almost 80 percent chose Smith. Heed the message here: Although both of the essays are *correct*, only one is *effective*.

It is not enough to write a "correct" essay or paper. Such writing often fails to advance your argument or accomplish the goal of your writing. Rather, strive for *effective* writing—writing that is clear, simple, precise, accurate, and concise. If your writing is effective, other people will not only enjoy reading your writing, but they will also think that you have a better-organized mind, are more competent, and, in this case, are a better biologist than someone who writes poorly. If readers can easily understand what you're saying, they are more likely to be impressed with and learn from your ideas.

Myths About Writing

Writing is judged by how easily it conveys ideas and helps us learn, not by its adherence to grammatical rules. To communicate well, you must understand what you did, what you want to say, and why it is important. You'll communicate effectively only when you have something to say, say what you mean, and logically support your statements with evidence. This requires no inspiration, wit, or rhythm. All you must do is make sure that your ideas are obvious to readers and never make readers guess or wonder if you have anything important to say.

---

[1]From Turk, Christopher, and John Kirkman. 1989 *Effective Writing*. 2nd edition. London: E. & F.N. Spon. For more information about this study, see Bardell, Ewa. 1978. "Does Style Influence Credibility and Esteem?" *The Communicator of Scientific and Technical Information* 35: 4–7; Turk, C.C.R. 1978. "Do You Write Impressively?" *Bulletin of the British Ecological Society* ix(3):5–10; Wales, LaRae H. 1979. *Technical Writing Style: Attitudes Towards Scientists and Their Writing.* Burlington, VT: University of Vermont Agricultural Experiment Station.

Many students assume that they are good writers, yet begrudgingly admit that they cannot distinguish a subject from a verb or a pronoun from a preposition. Others believe that they write well because they studied Latin, diagrammed a few thousand sentences, and can distinguish grammatical contraptions such as subjunctive pluperfect progressive and retained objects. Still others believe that good writing results from sincerity, from writing the way they speak, or from just being "gifted." The best evidence indicates that none of these things has much to do with one's ability to write well.

The first step toward improving one's writing involves understanding what writing is and what it isn't. Most conceptions about writing are, in fact, misconceptions.

Writing requires inspiration. Contrary to folklore, writing isn't sitting for hours in front of a blank piece of paper or computer screen, waiting to be "inspired." Writing never happens that way, not even for the best of writers. Writing only *looks* like it happens that way because many of the decisions that writers make are invisible to those who do not understand writing.

Writing is less an art than a craft. Like any craft, writing involves a series of decisions that, when done in correct order and with the proper attention to detail, can guarantee a decent and acceptable paper. That paper may not read as if it had been written by Hemingway, but it will communicate your thoughts effectively and help you learn about and understand your subject.

Writing to Learn Biology teaches practical and effective writing—writing that will communicate your message clearly and quickly. To help you understand what I mean, consider this sentence from the back of a tube of Crest toothpaste: "For best results, squeeze tube from the bottom and flatten it as you go up." No one would claim that this sentence is overly creative or that it represents great literature. However, it is perfect for its function because it efficiently does its job: It contains no excess words and its message cannot be misunderstood. Scientific writing, like this sentence on the back of a tube of toothpaste, is measured not by the pleasure that it gives, but by how well it does a job. Learning to write simple, functional sentences will help you learn about your subject and, in the process, make you a better student and a more productive professional. If you refuse to write simple, effective sentences, much of your education will be wasted because what you write will not be understood and therefore will be ignored.

The myth that effective writing occurs only when a writer is inspired is based on the notion that writing is something that happens to you when you're inspired. This is not true. Writing is something that you *do* —a process that helps you communicate and learn. Waiting to be inspired ensures failure because it postpones learning and justifies a writer's worst enemy: procrastination. Writing to learn requires only a few guidelines and a logical approach that are closer to common sense than to inspiration.

Just write the way you talk; after all, both writing and talking are means of communication. The fallacy of this approach is obvious when you read transcripts of conversations. Such transcripts seem disjointed and confusing without the gestures, pauses, facial expressions, and emphases that accompany talking. Read the transcript of a conversation and you'll see that you talk more loosely and informally than you would want to have recorded in writing. This is because talking is all that effective writing isn't: natural, informal, habitual, and relaxed. You've also practiced talking more than you have writing. Indeed, you speak more words in a month than you'll write during your lifetime.[2]

Good writers can quickly and effortlessly produce a perfect paper on their first try. This misconception typifies amateurs and poor writers. People who believe that good writing comes naturally on the first try are either incredibly talented or have low standards. For example, although the folksy stories of Andy Rooney and the homespun musings of Erma Bombeck seem so effortless and easy to understand, they require hours of hard work to write. Few people can let words flow without having them sound "spilled." As a philosopher said, "What is written without effort is read without pleasure."

Good writers know that good writing is hard work and that their first drafts are unsatisfactory. To write well, you must learn to revise well. As a famous writer once said, "There is no good writing, only good rewriting." *Writing to Learn Biology* will teach you how to revise your work—not as a means of merely correcting mistakes, but as a way of rethinking and learning more about your ideas, your subject, and your research.

Simple writing is not scholarly. People who use this excuse confuse simple writing with simplistic writing, and think that simple writing is not impressive or important. These people try to dignify their writing by choosing words such as "endeavor" instead of "try" because they think that endeavor is so much more, well, *sophisticated*. Such words are not sophisticated at all. They merely clutter your writing and bog readers down.

Contrary to what some people think, verbose and fancy writing usually hides shallow thought. More important, the research involving Smith's and Brown's essays (see p. 4) shows that a simple, straightforward style impresses readers more than the long-winded, "scholarly" writing style typical of many professionals. Heed the advice of the survey: You'll impress other people most not if you write in a scholarly manner but rather if you write so they can quickly and clearly understand what you're saying.

---

[2]On average, a person speaks about 12,000 words per day. This translates into an 84,000-word novel every week, and about 4 million words per year. Even Shakespeare, one of history's most prolific writers, did not write this much during his lifetime.

Good writing is an art. Effective writing is neither a science nor an art. Rather, it is a craft that can be learned by understanding and applying a few principles—not rules—of writing.

Effective writing is as simple as following a set of rules. Rules such as those for punctuation, spelling, and grammar are important to scientists and other professionals because they help them communicate their ideas. However, principles of grammar are not rigid rules, but rather are tools that help us detect and eliminate flaws in our writing that inhibit learning and communication. Although poorly written sentences and paragraphs are easy to identify because they sound awkward or confusing, correcting the problem is another matter entirely. This is best done by understanding the principles of writing, not by rattling off a list of grammatical rules. In many instances, a good ear will serve you better than a rigid rule.

Some rules are important. For example, a disregard for spelling and punctuation will make you seem uncaring, illiterate, or lazy. Other rules are useless to good writers (see "What do I need to know about grammar?" p. 144). Writers who rely only on rules may write "correctly" but not effectively. Rather than use writing as a tool to learn and communicate, these writers dwell on simplistic rules such as always using the passive voice, always stringing modifiers together, never splitting an infinitive, always repeating ideas, never using personal pronouns, and never beginning a sentence with a forbidden word. Such rules are panaceas that, without an understanding of writing, seldom improve one's writing or learning.

As shown by the essays of Smith and Brown, merely avoiding "errors" by following rigid rules does not ensure good writing. Moreover, pedantically correct writing is often dull and colorless. Rules for writing are valuable only if they improve communication. Use them as guidelines, not as substitutes for thought.

To write well you must have a large vocabulary. Words alone do not create meaning and communicate. If writing were that simple, then only a thesaurus or dictionary would be essential to becoming a good writer, and the size of one's vocabulary would determine one's writing skills. You don't need a huge vocabulary to write well. Indeed, a vocabulary of only about 1,000 words covers about 85 percent of a writer's needs to describe ordinary subjects. Thus, buying books, tapes, and videos with titles like "Nine Billion Gigantic Words That You Should Know" and "Super-Duper Word Power" that promise to make you a verbal Charlie Atlas won't improve your writing nearly as much as working to ensure that you write so that you cannot be misunderstood. Indeed, the words that you choose *not* to use are often more important than those that you use.

Reject the notion that long and unusual words are interesting and elegant. Elegance may be a by-product, but it can never be an intention

> **Science Headlines**
>
> Newspaper headlines are notorious for their double meanings. Articles about science are no exception, as shown by headlines such as "Milk Drinkers Are Turning To Powder" and "Scientists To Have Bush's Ear." Headlines such as "Genetic Engineering Splits Scientists" might make people think twice about a career in molecular biology, while other headlines describe astounding evolutionary tales. For example, an article entitled "Lung Cancer in Women Mushrooms" apparently describes some renegade female fungi that have evolved lungs and have started puffing on cigarettes. However, few will ever beat this gem that appeared in the travel section of a local newspaper: "Canada's Virgin Forests: Where The Hand Of Man Has Never Set Foot."

of your writing. As you write, try to learn to communicate rather than be clever or elegant. Take Albert Einstein's advice: "When you're out to describe the truth, leave elegance to the tailor."

Writing is less important than it used to be. Top scientists spend more than one-fourth of their working-day writing. Most of these scientists claim that their ability to write effectively helped advance their careers. Similarly, *Fortune* magazine recently asked successful executives what students should learn to help them prepare for careers. Their answer: *Learn to write well.* Editors and professors who must sift through heaps of poorly written papers would yell, "Amen."

Language determines how effectively you communicate. Although professionals who write poorly may fool readers for a paragraph or two, they eventually lose their readers.

Scientists' Excuses for Writing Poorly

Although scientists spend much of their time writing and may be highly educated, many express themselves wordily and obscurely, and therefore ineffectually. Listen to the laments of two editors of scientific journals:

> [Published papers] not only want of rhetorical finish (a slight blemish, comparatively speaking), but of all regard to correctness or appropriateness of language. . . . An inexcusable defect in composition, for the reason that it is so easily avoided, is the commonplace, inaccurate, in short, illiterate, language suffered to find its way into our journals.

> The majority of articles submitted for publication could be cut down one-half, and not a thought be eliminated in so doing. The repetition of well-known facts, padding with abstracts from text-books, and words, words, words, too often constitute the papers that appear as "original" in medical journals. And if the editor presumes to use the blue pencil in the least, the majority of authors consider it an insult.

Curiously, many scientists are eager to perpetuate the poor writing described by these editors. Rather than take responsibility for becoming a better writer and, therefore, a better scientist, these scientists rely on several excuses for their poor writing:

"Writing belongs in English classes, not science classes." This misconception underlies the notions that writing and science are unrelated, that words and science don't mix, and that writing is something done well only by people with names like Hemingway and Twain. All of these notions are incorrect.

Words and science are intimately linked, as are writing and thinking. Writing about biology forces you to *think* about biology—that's why writing is a such a powerful tool for learning.

"Great scientists write badly." This excuse is ridiculous. For example, Lewis Thomas and James Watson are biologists who have used personal, direct, and forceful writing to describe their brilliant ideas. Conversely, many other scientists' descriptions of research are as impersonal, dull, and lifeless as a phone-book or weather report.

Clear writing reveals how a clear mind attacks and solves a problem. Since one's writing reflects one's thinking, poor writers cannot pretend to be clear thinkers, much less effective scientists. In science, your ideas are only as good as your ability to express them to others.

"I just wasn't born a good writer. I'm a good scientist, not a good writer." This "blame nature" excuse typifies people who think that good science and effective writing are mutually exclusive. These individuals, who view scientific writing as a mystic art rather than a learnable craft, often become frustrated when their poor writing discredits their work. Contrary to what these people think, effective scientific writing is a craft that can be learned, not an effortless outpouring by a genius.

Poor writing reflects poor thinking: It is shifty and unpredictable, much like Brownian movement. Thus, most scientists who hide behind the "I'm a good scientist, not a good writer" excuse are usually only half right—right about not being a good writer, but usually wrong about being a good scientist.

"Simple writing is not scientific." Scientists who confuse simple writing with simplistic writing do not want to communicate clearly. They resist clear, simple writing because they know that clear, simple writing reveals problems and faulty logic, just as it can also reveal genius. Wrap the work in flowery, vague language, they reason, and others will ignore deficiencies in their work. Papers written by such scientists are usually ignored because other scientists cannot understand their message.

"The science is all that's important; if the science is good, then blemishes in the writing are irrelevant." Scientists with this "grammar don't matter" attitude either have an inflated self-opinion or understand little

about science. Writing is an integral part of thinking, learning, and science. Data cannot stand alone. Rather, they must be interpreted and incorporated into an argument supporting or refuting a hypothesis. "Blemishes" such as ambiguity, poor grammar, unnecessary jargon, and inaccuracy typify shoddy thinking and carelessness. These blemishes produce weak arguments that discredit your work.

Scientists who think that what matters is what you say, not how you say it, are half right. Trite, shallow, or illogical ideas do not magically become brilliant ideas when they're written well: An embroidered sow's ear remains a sow's ear, and style is no substitute for substance. However, neither can substance substitute for a lack of style. Ideas buried in confusing writing aren't worth much because they're nothing more than confusion. Knowing how to use writing to learn will help identify your weak ideas, thereby allowing you to strengthen them as you learn more about your ideas and your subject.

Scientific writers establish their credibility in two ways: (1) with facts, evidence, and logic and (2) with their writing style. Although poor writing won't camouflage a lack of substance, good ideas are useless unless they're explained well. The hallmark of important research papers is that they're readily understandable to other scientists. Scientists naive enough to believe that writing is irrelevant to science are often equally naive about research.

"Don't worry about the details." The effectiveness of writing depends on details, which, if ignored, can damage your credibility. For example, one student submitted a paper describing a nonexistent and apparently satanic chemical having the formula $C_6H_{12}O_{666}$. Similarly, I recently received an advertisement for a seminar entitled, "Effective Scientific Writting." As best I could tell, the only consolation was that the course was non-credit.

Using words carelessly can confuse readers, as can mistakes created by "details" such as misplaced decimals, incorrect measurements, or grammatical mistakes such as stacked modifiers. Biologists who ignore these details are often ineffective scientists because their work is either incomprehensible or unrepeatable and is therefore ignored. Sloppy writing can have even more serious implications. For example, the "Code for Communications" of the Society for Technical Communication states that precision, use of simple and direct language, responsibility for how readers understand your message, and respect for readers' needs for information (rather than your need for self-expression) show a technical writer's commitment to professional excellence and *ethical* behavior. "Details" that produce vague, misleading papers damage your credibility and hinder science.

"They'll know what I mean." This "they can read my mind" excuse for poor writing tries to shift the responsibility for communicating to the reader by forcing the reader to do your thinking for you. This approach

fails because it is always the writer's job to communicate with the reader. Papers must communicate, not merely "make information available to others," and scientific papers must be read *and understood*. Merely converting your data and observations to sentences is irrelevant; it matters only that readers accurately perceive what you had in mind. This requires no fancy words, grammatical tricks, or gimmicks. Rather than impress readers, these ornaments usually only confuse. Effective scientific writing has no place for fanciful leaps or implied truths. Facts and deductions are the rule.

Regardless of its difficulty or sophistication, a complex subject can be made as accessible as a simple subject by an effective writer. Never mind what you think your writing is *supposed* to mean—all that matters is what it *says*. Readers will know what you say, but only if you write effectively will they know what you *mean*. How you write about science *is* the science. Similarly, your ideas alone, no matter how brilliant, will be irrelevant if you can't describe them effectively. As Vladimir Nabokov wrote, "Style and structure are the essence of a book; great ideas are hogwash."

"Scientists have always written in scholarly prose." This "we've always done it this way" excuse is used by many poor writers. These writers, if describing atomic structure, would write something like this:

> It is hypothesized by this author that, in essence, the initial material existence of physically manifest substances was relatively dense, massive in weight, durable, and particulate in form; the extreme manifestation of hardness being categorically displayed by resistance to diminution in size due to abrasive processes and by counter-fragmentation systems.

Although such writing is common in many scientific journals, scientists have not always written like this. The sentence you just read is, in fact, a "modern" version of a sentence written by Isaac Newton almost 300 years ago:

> It seems probable to me, that God in the beginning formed matter in solid, massy, hard, impenetrable, moveable particles; . . . even so very hard, as never to wear or break in pieces.[3]

The scientific literature contains many other examples of poetic and aesthetic responses to scientific experiments. For example, the Roman philosopher Lucretius wrote *De Rerum Natura (On the Nature of Things),* which summarized scientific thought in his time, as a poem. In 1665 Robert Hooke wrote, "These pleasing and lovely colours have I also sometimes with pleasure observ'd even in Muscles and Tendons."[4]

[3]Newton, Sir Isaac. 1704. *Optiks.* London. Quoted by Pyke, Magnus. 1960. In: *This Scientific Babel.* The Listener, LXIV (1641)(8th September), 380.
[4]Espinasse, M. 1956. *Robert Hooke.* London: Heinemann.

Similarly, you quickly sense Malpighi's excitement when he quotes Homer in a letter written upon seeing capillaries for the first time: "I see with my own eyes a certain great thing."[5]

"Scholarly" writing often impedes communication because it unnecessarily forces readers to wade through clutter such as *it is hypothesized by this author*. Here's a biological example of scholarly writing.

> The following describes the activities of five immature mammals of the family of nonruminant artiodactyl ungulates. All five of these may be described as being of less than average magnitude; however, no information is given as to the relative size of one with respect to another. Available evidence indicates that the first of the group proceeded in the direction of an area previously established for the purpose of commerce. Data on the second of the group clearly show that, at least during the time period under consideration, it remained within the confines of its own place of residence. Reports received on the activities of the third member of the group seem to show conclusively that it possessed an unknown quantity of the flesh of a bovine animal, prepared for consumption by exposure to dry heat. The only information available on the fourth member of the group is of a wholly negative nature, namely, that its possessions did not include any of the type previously described as having been in the possession of its predecessor in this discussion. As to the fifth and last member of the group, fairly conclusive evidence points to its having made, during the entire course of a movement in the direction of its place of residence, a noise described as "wee, wee, wee."[6]

As you discovered at the end of this paragraph, this is a "scholarly" version of "This Little Pig Went to Market." Although you may have enjoyed hearing this nursery rhyme as a child while your smiling mother or father pulled at your toes, you probably found this scholarly version annoying—annoying because its most noticeable trait is its writing, not its message.[7]

Good writing is invisible because it draws attention not to itself, but to the writer's ideas. Readers don't automatically think, "This sure is good grammar" when they read a well-written paper. Rather, they quickly and

---

[5]Marcello Malpighi was an Italian physician who founded microscopic anatomy, demonstrating that blood flows to tissues through tiny capillaries too small to be seen with the naked eye. William Harvey, a famous English physician who discovered the circulation of blood, had inferred that these capillaries must exist, but he had never seen them. See Foster, M. 1901. *Lectures on the History of Physiology during the Sixteenth, Seventeenth and Eighteenth Centuries*. Cambridge: Cambridge University Press.

[6]From Mancuso, Joseph C. 1990. *Mastering Technical Writing*. Reading, MA: Addison-Wesley Publishing Co.

[7]Tired, prefabricated phrases such as *at this point in time, pursuant to our agreement*, and *your assistance in this matter will be greatly appreciated*, common in "professional" writing, are similar to the dreaded cousin who tells the same stories over and over again regardless of the occasion. Nevertheless, many people consider such phrases an important aspect of sounding "professional." These people should reconsider. Indeed, the *Harvard Business Review* studied 800 letters written by the most prominent business executives in the United States and concluded that *all* of these letters were concise and clear—almost the exact opposite of what many call "professional."

clearly understand the author's message. Similarly, the mistakes that typify poor writing do not announce themselves with titles such as "dangling modifier" or "passive voice." Readers know the writing is poor because it doesn't communicate clearly or quickly.

Reject excuses for poor writing. Do not let them betray your ideas and ability to think.

Why is Writing Important?

If any man wishes to write a clear style, let him first
be clear in his thoughts.
    —Johann W. von Goethe

A good narrative style does not attract undue
attention to itself. Its job is to keep the reader's mind
on the story, on what's happening, the event, and
not the writer.
    —Leon Surmelian

Speech is the representation of the mind, and writing
is the representation of speech.
    —Aristotle

Effective writing is extremely rewarding, but it can be hard and lonely work. Words seldom flow, and writing rarely comes easily because effective writing requires sustained thought. Clear thinking, like clear writing, is a conscious act that biologists and other professionals must force upon themselves. It's also critical to your career, both as a student and as a professional:

- Learning to write effectively will improve your grades. Much research shows that students who write well make better grades than do students who write poorly.

- Effective writing saves time and money. Boring writing is tossed aside unread, a waste of the investment made to produce it. Similarly, costly research must be repeated when scientists do not clearly describe their methods and results.

- Regardless of whether you become a scientist, businessperson, lawyer, or other professional, your ability to write will help determine the impact of your ideas and work. Many brilliant ideas have died in obscurity, their discoverers unknown because they wrote leaden prose that no one could or wanted to understand. Since these ideas were not communicated well, they were overlooked and ignored. Conversely, the recognition that accompanies well-written papers can lead to promotions, awards, and salary raises. Heed the advice of Sir William Osler: "In science the credit goes to the man who convinces the world, not to the man to whom the idea first comes." To do yourself justice, you must think and write well.

- Writing and thinking exert a back-pressure on each other that greatly enhances learning. Writing about a subject forces you to be precise. In doing so, writing helps you uncover faulty logic that betrays your search for truth and knowledge.

  Careful writing helps you develop ideas and therefore is an important tool for helping you to learn (such as "taking notes"), observe (such as with a sketch), plan, show relationships, review, organize, communicate, remember, clarify, and discover what you know, what you don't know, and what you need to know. We think and capture our thoughts with words. Therefore, consistently poor writing betrays one's inability or unwillingness to think clearly. It is more than stylistic inelegance; it's an outward and visible sign of inner confusion.

  Writing and thinking are related tasks. That's why nothing helps you learn about and clarify a topic better than writing. To understand this concept, consider the following paragraph questioning if a law requiring a five-cent deposit on bottles and cans reduces litter:

> The law wants people to return the bottles for five cents, instead of littering them. But I don't think five cents is enough nowadays to get people to bother. But wait, it isn't just five cents at a blow, because people can accumulate cases of bottles or bags of cans in their basements and take them back all at once, so probably they would do that. Still, those probably aren't the bottles and cans that get littered anyway; it's the people out on picnics or kids hanging around the streets and parks that litter bottles and cans, and they sure wouldn't bother to return them for a nickel. But someone else might—Boy Scout and Girl Scout troops and other community organizations very likely would collect the bottles and cans as a combined community service and fund-raising venture; I know they do that sort of thing. So litter would be reduced.[8]

This first draft, besides illustrating everyday reasoning, exemplifies the close relationship between thinking and writing. Each sentence challenges the preceding one and, in the process, advances the writer's argument.

Writing is not a dull, irrelevant task to be endured. Rather, it will enhance your career by helping you to learn and communicate. Many of your best ideas will be lost to you if you do not write well.

- In many professions, writing is the primary means by which you will become known (or remain unknown). This is especially true in science. Data cannot speak for themselves; rather, they must be interpreted and

---

[8]Perkins, David N., R. Allen, and J. Hafner, 1983. "Difficulties in Everyday Reasoning," In *Thinking: The Expanding Frontier*, ed. W. Maxwell. Hillsdale, NJ: Lawrence Erlbaum Associates.

used to make an argument supporting or rejecting a hypothesis. Effective writing helps you advance your argument and is therefore the primary way for you to establish your reputation. You must use written evidence to convince your peers that your hypotheses are significant and that your ideas are good. Indeed, neither dexterity in the lab, innate knowledge, familiarity with the literature, brilliant ideas, creativity, nor personal charm will overcome the frustration and poor reputation established by poor writing. Poor writing breeds distrust and prompts readers to wonder if language is the writer's only area of incompetence.

The claims of science are the products of persuasion, for only through persuasion can you establish the importance and meaning of your work. Don't let poor writing hide the importance and meaning of your ideas.

• Science is a collective enterprise based on the free exchange of ideas. Consequently, writing is critical to science because it extends the history of science and is the vehicle for sharing scientific knowledge. Scientists must publish their work in a permanent and retrievable form so that others can examine, test, and build on their data and ideas. Publications are the cornerstone of science because they confirm or add to our knowledge.

• The goal of scientific research is to discover and communicate new information. Communicating new information requires clear, concise writing to convince colleagues of the importance and validity of your discovery. Consequently, your ability to write effectively will largely determine your influence as a scientist. Indeed, merely "doing research," having a breakthrough idea, designing a brilliant experiment, or making an important discovery will do you little good if you cannot communicate its brilliance or importance to others. Stated simply, scientists must do *and write about* science. Research is incomplete until it is published and made available for other scientists to test and build on. Writing that is verbose, pretentious, and dull communicates poorly, delays or prevents publication, and therefore delays scientific progress.

• Biologists at most "major" universities must publish their research in peer-reviewed journals to keep their jobs. This "publish or perish" requirement means that biologists unable to write effectively seldom keep their jobs (see "Publish or Perish," p. 240). Similarly, most biologists must obtain money from agencies such as the National Science Foundation and the National Institutes of Health to support their research. Getting this money requires that biologists write research proposals. The inability of biologists to write convincingly almost always means that they will get no money for their work, and therefore that they cannot do the research. This results in no data for publication which, in turn, often causes the biologists to be fired.

Scientists who write well often progress faster in their careers than do

Keeping a Personal Journal

> I had, also, during many years followed a golden rule, namely, that whenever a published record, new observation or thought came across me, which was opposed to my general results, to make a memorandum of it without fail and at once; for I had found by experience that such facts and thoughts were far more apt to escape from the memory than favourable ones. — Charles Darwin, *Life and Letters*

> In my journal, anyone can make a fool of himself. — Rudolph Virchow

Most serious thinkers, including many biologists, keep an informal journal to capture their thoughts. Indeed, the journals of naturalists and scientists such as Thoreau, Darwin, Freud, and Einstein mapped the birth and development of their revolutionary ideas. This highlights a journal's primary value: discovery. Such informal writing helps you sort your ideas and theories and, in the process, discover new ideas. As discussed in Chapter 2, such free-writing often takes you where you never intended to go — not by reflecting your ideas, but by *leading your thoughts*. There are few better ways to learn about yourself and your thoughts than by keeping a journal, because writing about what you don't know is an excellent way to know. In pausing to write, you'll listen to yourself think.

Journals are a cross between a student notebook and writer's diary and describe topics such as "What happened in class today?" and "Why do I believe this?" Journals may include observations about a field site, speculation about data, and syntheses of your ideas. Personal journals are also excellent places to voice questions and express your doubts. This is important because writing questions often helps sharpen a question so that it can be answered.

Keep a personal journal. Since no one will see the journal but you, feel free to write about whatever you please. After a while, you'll treasure the time you set aside for writing in your journal. It will become a relaxing time of discovery.

those who write poorly. Thus, knowing how to write effectively will help you get and keep a job and is therefore an investment in your future. Also, since the printed word in indelible, you must write well to protect your future reputation.

• Finally, scientists must communicate their discoveries to the public. Einstein said it best:

> It is of great importance that the general public be given the opportunity to experience — consciously and intelligently — the efforts and results of scientific research. It is not sufficient that each result be taken up, elaborated, and applied by a few specialists in the field. Restricting the body of knowledge to a small group deadens the philosophical spirit of a people and leads to spiritual poverty.

Your ability to write effectively will, in part, determine your success as a professional. Despite the importance of writing, most students are trained only to obtain information rather than to communicate it. Consequently, most students studying science learn "scientific writing" by imitating their mentors or colleagues. Indeed, scientific writing is a learned "skill," as described by Michael Crichton in the *New England Journal of Medicine*:[9]

> It now appears that obligatory obfuscation is a firm tradition within the medical profession . . . [Medical writing] is a highly skilled, calculated attempt to confuse the reader. . . . A doctor feels he might get passed over for an assistant professorship because he wrote his papers too clearly—because he made his ideas seem too simple.

Consequently, students often perpetuate the pompous, boring, and ineffective writing that has characterized many scientific journals for decades. In many instances, it sounds as if scientists are trying to keep secrets rather than communicate.

In the following chapters, you'll learn about the process of writing and the tools that good writers use to discover and communicate their ideas. However, knowing about these tools is about as useful as listing just the ingredients of a great recipe and then expecting someone to make the dish. Just as knowing the ingredients and how to use them distinguishes reading cookbooks from cooking, how you use what you learn in this book will distinguish reading this book about writing from writing. Practice what you learn.

---

[9]Michael Crichton is a physician and professional writer. His first novel, the suspenseful *The Andromeda Strain*, centered on complex scientific issues and became a best-seller, prompting him to devote increasingly more time to writing. His other books include *Sphere* and *Jurassic Park*.

Lewis Thomas

Germs

Lewis Thomas (b. 1913) is a research physician who
has studied hypersensitivity, the pathogenicity of
mycoplasmas, and infectious diseases. Most of his
essays were originally published in *The New
England Journal of Medicine* and later in books.
The first of these books was *The Lives of a Cell:
Notes of a Biology Watcher*, which won the
National Book Award in 1975.

Watching television, you'd think we lived at bay, in total jeopardy,
surrounded on all sides by human-seeking germs, shielded against
infection and death only by a chemical technology that enables us to
keep killing them off. We are instructed to spray disinfectants
everywhere, into the air of our bedrooms and kitchens and with special
energy into bathrooms, since it is our very own germs that seem the
worst kind. We explode clouds of aerosol, mixed for good luck with
deodorants, into our noses, mouths, underarms, privileged crannies —
even into the intimate insides of our telephones. We apply potent
antibiotics to minor scratches and seal them with plastic. Plastic is the
new protector; we wrap the already plastic tumblers of hotels in more
plastic, and seal the toilet seats like state secrets after irradiating them
with ultraviolet light. We live in a world where the microbes are always
trying to get at us, to tear us cell from cell, and we only stay alive and
whole through diligence and fear.

We still think of human disease as the work of an organized,
modernized kind of demonology, in which the bacteria are the most
visible and centrally placed of our adversaries. We assume that they must
somehow relish what they do. They come after us for profit, and there
are so many of them that disease seems inevitable, a natural part of the
human condition; if we succeed in eliminating one kind of disease there
will always be a new one at hand, waiting to take its place.

These are paranoid delusions on a societal scale, explainable in part
by our need for enemies, and in part by our memory of what things
used to be like. Until a few decades ago, bacteria were a genuine
household threat, and although most of us survived them, we were
always aware of the nearness of death. We moved, with our families, in
and out of death. We had lobar pneumonia, meningococcal meningitis,

streptococcal infections, diphtheria, endocarditis, enteric fevers, various septicemias, syphilis, and, always, everywhere, tuberculosis. Most of these have now left most of us, thanks to antibiotics, plumbing, civilization, and money, but we remember.

In real life, however, even in our worst circumstances we have always been a relatively minor interest of the vast microbial world. Pathogenicity is not the rule. Indeed, it occurs so infrequently and involves such a relatively small number of species, considering the huge population of bacteria on the earth, that it has a freakish aspect. Disease usually results from inconclusive negotiations for symbiosis, an overstepping of the line by one side or the other, a biologic misinterpretation of borders.

Some bacteria are only harmful to us when they make exotoxins, and they only do this when they are, in a sense, diseased themselves. The toxins of diphtheria bacilli and streptococci are produced when the organisms have been infected by bacteriophage; it is the virus that provides the code for toxin. Uninfected bacteria are uninformed. When we catch diphtheria it is a virus infection, but not of us. Our involvement is not that of an adversary in a straightforward game, but more like blundering into someone else's accident.

I can think of a few microorganisms, possibly the tubercle bacillus, the syphilis spirochete, the malarial parasite, and a few others, that have a selective advantage in their ability to infect human beings, but there is nothing to be gained, in an evolutionary sense, by the capacity to cause illness or death. Pathogenicity may be something of a disadvantage for most microbes, carrying lethal risks more frightening to them than to us. The man who catches a meningococcus is in considerably less danger for his life, even without chemotherapy, than meningococci with the bad luck to catch a man. Most meningococci have the sense to stay out on the surface, in the rhinopharynx. During epidemics this is where they are to be found in the majority of the host population, and it generally goes well. It is only in the unaccountable minority, the "cases," that the line is crossed, and then there is the devil to pay on both sides, but most of all for the meningococci.

Staphylococci live all over us, and seem to have adapted to conditions in our skin that are uncongenial to most other bacteria. When you count them up, and us, it is remarkable how little trouble we have with the relation. Only a few of us are plagued by boils, and we can blame a large part of the destruction of tissues on the zeal of our own leukocytes. Hemolytic streptococci are among our closest intimates, even to the extent of sharing antigens with the membranes of our muscle cells; it is our reaction to their presence, in the form of rheumatic fever, that gets us into trouble. We can carry brucella for long periods in the cells of our reticuloendothelial system without any awareness of their existence; then cyclically, for reasons not understood but probably related to immunologic reactions on our part, we sense them, and the reaction of sensing is the clinical disease.

Most bacteria are totally preoccupied with browsing, altering the configurations of organic molecules so that they become usable for the energy needs of other forms of life. They are, by and large, indispensable to each other, living in interdependent communities in the soil or sea. Some have become symbionts in more specialized, local relations, living as working parts in the tissues of higher organisms. The root nodules of legumes would have neither form nor function without the masses of rhizobial bacteria swarming into root hairs, incorporating themselves with such intimacy that only an electron microscope can detect which membranes are bacterial and which plant. Insects have colonies of bacteria, the mycetocytes, living in them like little glands, doing heaven knows what but being essential. The microfloras of animal intestinal tracts are part of the nutritional system. And then, of course, there are the mitochondria and chloroplasts, permanent residents in everything.

The microorganisms that seem to have it in for us in the worst way—the ones that really appear to wish us ill—turn out on close examination to be rather more like bystanders, strays, strangers in from the cold. They will invade and replicate if given the chance, and some of them will get into our deepest tissues and set forth in the blood, but it is our response to their presence that makes the disease. Our arsenals for fighting off bacteria are so powerful, and involve so many different defense mechanisms, that we are in more danger from them than from the invaders. We live in the midst of explosive devices; we are mined.

It is the information carried by the bacteria that we cannot abide.

The gram-negative bacteria are the best examples of this. They display lipopolysaccharide endotoxin in their walls, and these macromolecules are read by our tissues as the very worst of bad news. When we sense lipopolysaccharide, we are likely to turn on every defense at our disposal; we will bomb, defoliate, blockade, seal off, and destroy all the tissues in the area. Leukocytes become more actively phagocytic, release lysosomal enzymes, turn sticky, and aggregate together in dense masses, occluding capillaries and shutting off the blood supply. Complement is switched on at the right point in its sequence to release chemotactic signals, calling in leukocytes from everywhere. Vessels become hyperreactive to epinephrine so that physiologic concentrations suddenly possess necrotizing properties. Pyrogen is released from leukocytes, adding fever to hemorrhage, necrosis, and shock. It is a shambles.

All of this seems unnecessary, panic-driven. There is nothing intrinsically poisonous about endotoxin, but it must look awful, or feel awful when sensed by cells. Cells believe that it signifies the presence of gram-negative bacteria, and they will stop at nothing to avoid this threat.

I used to think that only the most highly developed, civilized animals could be fooled in this way, but it is not so. The horseshoe crab is a primitive fossil of a beast, ancient and uncitified, but he is just as vulnerable to disorganization by endotoxin as a rabbit or a man. Bang has shown that an injection of a very small dose into the body cavity will cause the aggregation of hemocytes in ponderous, immovable masses that block the vascular channels, and a gelatinous clot brings the circulation to a standstill. It is now known that a limulus clotting system, perhaps ancestral to ours, is centrally involved in the reaction. Extracts of the hemocytes can be made to jell by adding extremely small amounts of endotoxin. The self-disintegration of the whole animal that follows a systemic injection can be interpreted as a well-intentioned but lethal error. The mechanism is itself quite a good one, when used with precision and restraint, admirably designed for coping with intrusion by a single bacterium: the hemocyte would be attracted to the site, extrude the coagulable protein, the microorganism would be entrapped and immobilized, and the thing would be finished. It is when confronted by the overwhelming signal of free molecules of endotoxin, evoking memories of vibrios in great numbers, that the limbulus flies into panic, launches all his defenses at once, and destroys himself.

It is, basically, a response to propaganda, something like the panic-producing pheremones that slave-taking ants release to disorganize the colonies of their prey.

I think it likely that many of our diseases work in this way. Sometimes, the mechanisms used for overkill are immunologic, but often, as in the limulus model, they are more primitive kinds of memory. We tear ourselves to pieces because of symbols, and we are more vulnerable to this than to any host of predators. We are, in effect, at the mercy of our own Pentagons, most of the time.

How does Thomas achieve such a graceful style?

Thomas has been described as a scientist with the mind of a humanist. Based on this essay, do you agree? Why or why not?

Isaac Asimov

The Next Frontiers For Science

Isaac Asimov (b. 1920) is a biochemist and world authority on science and medicine. He's world-famous for his science-fiction books and articles. He's incredibly prolific: He's written more than 450 books.

His essay entitled "The Next Frontiers For Science" was published in 1991 in several popular magazines, including *Money* and *Fortune*. He makes a controversial claim—that the only scientific research that deserves major funding is research to save our planet.

Space Exploration? Genetic Research? The Environment? A Hard Choice Must be Made.

We are living in an age where many scientists are thinking big. There is the supercollider, a new unprecedentedly powerful particle accelerator which may give us an answer at last to the final details of the structure of the universe, its beginning, and its end.

There is the genome project, which will attempt to pinpoint every last gene in the human cells and learn just exactly how the chemistry of human life (and of inborn disease) is organized.

There is the space station, which will attempt, at last, to allow us to organize the exploitation of near space by human beings.

All these things, and others of the sort, are highly dramatic and will be, at least potentially, highly useful. All are also highly expensive, something of great importance in a shrinking economy. Worse yet, all are, at the moment, highly irrelevant.

What is relevant is that we are destroying our planet.

A steadily increasing population is placing ever-higher demands on Earth's resources, is forcing the conversion of more and more land to human needs and is wiping out the wilderness and the ecological balance of the planet, something on which we all depend.

A steadily rising use of fossil fuels for energy (at a rate that is increasing more rapidly than the population) is choking Earth's atmosphere with gases that are slowly poisoning it. In addition, it allows the atmosphere to conserve heat more efficiently, so that the planet is experiencing a greenhouse effect that may have catastrophic impact.

A steadily increasing production of chemical substances that are highly toxic, or that cannot be recycled by biological processes, or both, is poisoning the soil and water of the Earth, is destroying the ozone layer and is converting much of the planetary surface into a garbage heap. Since there can be nothing on Earth, simply nothing, that is more important than saving the planet, our coming priorities must be to reverse these destructive tendencies. And America must lead. It is, for instance, foolish, absolutely foolish, to put the study of reproductive physiology to work on test-tube babies and on producing babies after menopause; we must not increase the number of babies, but decrease them.

We must find alternate sources of energy, long-lasting and non-polluting. We must continue the search for nuclear fusion, in the hope that it will be a far richer and safer source than nuclear fission. We must develop wind-power, wave-power, the use of Earth's internal heat and, most of all, the direct use of solar power. All these things are highly practical, but cost more money than oil and coal, so the challenge is to make them cheaper. (The fact that we can destroy our planet so cheaply, by the way, does not mean we ought to destroy it.)

We must find ways of detoxifying toxic products produced by industrial plants. We must find substitutes for packaging, substitutes that are recyclable. We must find substitutes for chemicals that destroy the ozone layer.

We must find methods of saving our forests, of saving threatened species, of maintaining a healthy ecological balance on Earth.

If there is any spare effort left over from these absolute necessities of scientific advance, we can put them into other projects — otherwise not.

I regret this, for I am emotionally on the side of the big projects, all of them, but necessity is a hard task-master, and necessity is now in the saddle and holds the whip.

Do you agree with Asimov? Write a one-page essay that describes your opinion. Substantiate your opinion with evidence, and develop your ideas with logic.

The Wilder Side of Writing About Biology

Every year we spend billions of dollars on biological research. While many of these studies have described important discoveries such as cures for diseases, others have described leaps into the absurd. For example, a prestigious medical journal included this description of bringing mice back from the dead.

> The extract was injected into several mice, which died within two to six minutes. Given smaller doses, the mice recovered.

Another article described how mice were transformed into a drug:

> The mice were treated with penicillin, then changed to erythromycin.

The results of similar studies are described in William Hartston's book entitled *Drunken Goldfish & Other Irrelevant Scientific Research* (published in 1987 by Fawcett Crest, New York). For example, scientists have published papers documenting that alcohol can help cure premature ejaculation in neurotic dogs, that rats are more attracted to other rats than to tennis balls, and that holy water does not affect the growth of radishes. Other studies have determined whether goldfish have hangovers, while still others have involved feeding blood of schizophrenics to spiders.

As you might guess, sex is a favorite subject of much biological research. One unusual paper described kinky sex practices that improve a male goat's performance, while another, entitled "Paroxysmal Sneezing Following Orgasm" and published in the *Journal of the American Medical Association*, suggests that it might be a good idea for some folks to keep a box of hankies next to the bed.* However, perhaps the most intriguing title of a paper appears in *Veterinary Record:* "The Erect Dog Penis: A Paradox of Flexible Rigidity." Rather than spoil your fun, I'll let you look up that one for yourself.

*In case you're wondering, the recommended treatment for this problem is applying a long-acting nasal decongestant to each nostril about 30 minutes before intercourse. For more laughs about science and scientists, read the *Journal of Irreproducible Results.* Instead of tediously asking, "Is it scientific?," this parody of scientific journals asks, "Is it funny?" My favorites in the journal include "Bovinity," "The Inheritance Pattern of Death," "Therapeutic Effects of Forceful Goosing on Major Affective Illness," "Reading Education for Zoo Animals: A Critical Need," "Notes Upon A Whopper," "A Drastic Cost-Saving Approach to Using Your Neighbor's Electron Microscope," and "Weekend Science: Let's Make a Thermonuclear Device."

Exercises

1. Science greatly affects our lives, yet the public knows little about scientists. Indeed, many publications stereotype scientists as soulless, eccentric hermits studying things that they can't explain in anything except gibberish. Similarly, movies such as *Back to the Future III, Gremlins 2: The New Batch, Die Hard 2, Total Recall,* and *E.T.* depict scientists as heartless and morally blind—determined to dissect or otherwise abuse lovable creatures such as a beautiful mermaid, E.T., a baby dinosaur, a revived neolithic man, and intelligent chimpanzees.

What are your conceptions of scientists? Nobel laureate James Watson said, "One could not be a successful scientist without realizing that, in contrast to the popular conception supported by newspapers and mothers of scientists, a goodly number of scientists are not only narrow-minded and dull, but also just stupid." Do you agree with this statement? Why or why not?

2. Several years ago the U.S. Supreme Court upheld biologists' right to patent new forms of life. The vote on this case was 5–4, indicating that the justices' opinions were divided. How would you have ruled on the case? Write an essay explaining your position. Your argument is directed at people who hold the opposing view, and your purpose is to persuade them that your opinion is correct.

3. Choose one of the articles in the "Medicine" section of one of the eight most recent issues of *Time* magazine. Summarize the article for a friend who has not read it. Do not evaluate the article or give your opinion about it; simply inform your friend of its contents.

4. List what you feel are the four most important ideas in biology. Include a brief discussion of each idea and of your ranking. Compare your ranking with that of Robert Hazen and James Trefil in *Science Matters: Achieving Scientific Literacy:*

All living things are made from cells, the chemical factories of life.

All life is based on the same genetic code.

All forms of life evolved by natural selection.

All life is connected.

Do you agree with their ranking? Why or why not?

5. Find and read a short article in a journal or magazine that you find easy to read and understand. Study the writing—why is the article easy to read? How many of the sentences are simple? Are there any grammatical mistakes? Spelling mistakes? How do the paragraphs lead you to the conclusion?

6. Write a one-page opinion on these subjects:

Governmental regulation of population growth.

Mandatory recycling.

Genetic engineering.

Include three facts to support each opinion. Then write a one-page opinion that opposes the one you just wrote (you may not agree with these opinions, but write them anyway). Include two facts to support these opinions.

7. How does biology affect your life? Write for 10 minutes about your favorite topic in biology. What questions did you think of as you wrote?

8. Much of what biologists have written has generated controversy. For example, Darwin devoted only one paragraph of *The Origin of Species* to humans, yet that paragraph was a focal point of controversies that followed:

In the future I see open fields for far more important researches. Psychology will be securely based on the foundation already well laid by Mr. Herbert

Spencer, that of the necessary acquirement of each mental power and capacity by gradation. Much light will be thrown on the origin of man and his history.

Why was this controversial? Do you agree with Darwin? Why or why not?

9. On the desk of John Updike (a two-time Pulitzer Prize winner) is a framed copy of a quotation of Pliny the Elder: "Nulla dies sine linea" ("Not a day without a line"). Why do you think Updike has this quotation on his desk? How does this relate to the myth of writers needing "inspiration" to write effectively?

10. Write a one-page essay about any of the following topics:

    The structure of your body.

    How you digest pizza.

    Why you might faint if you stand at attention too long.

    Why white blood cells are similar to the defenders of a fort.

    Why someone should quit smoking.

11. Do you agree with the idea that clear writing leads to clear thinking? Cite examples of how putting your thoughts down on paper has helped clarify your ideas.

12. Support or refute this statement:

    Having knowledge is not what makes a scientist. For example, if all of the scientific information known today were to be accepted with no new work, there would be no science. Statements such as "All cells come from preexisting cells" may look scientific, but they are not science if the only reason you believe them is because you read them or because your professor said them.

13. Many people base their lives on blind faith in things such as astrology, fortunetellers, and religious dogma. Science has also had its share of blind faith. For example, the Scholastics of the late Middle Ages accepted Aristotle's conclusions merely because Aristotle had said them, not because the conclusions were supported by data. Does blind faith have a place in science? If so, when and where? If not, why?

14. Find an example of scientific writing in a newspaper or magazine. Why do you consider the article scientific? What was the purpose of the article?

15. Write a letter to the editor of your local newspaper expressing your view or concern about a scientific topic.

16. Discuss, support, or refute the ideas included in these quotations:

    Ants are so much like human beings as to be an embarrassment. They farm fungi, raise aphids as livestock, launch armies into wars, use chemical sprays to alarm and confuse enemies, capture slaves. . . . They exchange information ceaselessly. They do everything but watch television. —Lewis Thomas

    Nature has no more concern or praise for human souls than for ants. —Giacomo Leopardi

The main object of all science is the freedom and happiness of man. — Thomas Jefferson

It is not what the man of science believes that distinguishes him, but how and why he believes it. — Bertrand Russell

Circumstantial evidence can be overwhelming. We have never seen an atom, but we nevertheless know that it must exist. — Isaac Asimov

Learning and memory are just as much biology as a process of DNA replication. — Anne A. Ehrhardt

Evolution is more important than living. — Ernst Jünger

It's humbling to think that all animals, including human beings, are parasites of the plant world. — Isaac Asimov

There is nothing in which the birds differ more from man than the way in which they build and yet leave a landscape as it was before. — Robert Staughton Lynd

Eventually, we'll realize that if we destroy the ecosystem, we destroy ourselves. — Jonas Salk

Today, the theory of evolution is an accepted fact for everyone but a fundamentalist minority, whose objections are based not on reasoning but on doctrinaire adherence to religious principles. — James D. Watson

It is the easiest thing in the world to deny a fact. People do it all the time. Yet it remains a fact just the same. — Isaac Asimov

17. Why do you write? What kinds of writing do you do? What is the purpose of each kind of writing? What kind do you like best? What kinds do you dislike? Why?

18. What kinds of publications do you read? What has this reading taught you in the last week? What effects does it have on you and what you do?

19. Write a one-page essay about what you think of biology and your biology class.

20. What is effective writing? Do you write well? What evidence do you have to support your answer?

21. Support or refute this statement: Truths of science are the achievements of arguments.

22. Write a short essay describing how this *The Far Side* cartoon relates to biology:

THE FAR SIDE By GARY LARSON

"Hold it right there, young lady! Before you
go out, you take off some of that makeup and
wash off that gallon of pheromones!"

Figure 1 – 1
The Far Side. Copyright 1988. UNIVERSAL PRESS SYNDICATE. Reprinted with permission.

CHAPTER TWO

Preparing a First Draft:

Writing to Learn

Writing and thinking and learning [are] the same process.
    —William Zinsser, *Writing to Learn*

The improvement of understanding is for two ends: first our own increase of knowledge; secondly to enable us to deliver that knowledge to others.
    —John Locke

Thinking is the activity I love best, and writing to me is simply thinking through my fingers.
    —Isaac Asimov

. . . Making mental connections is our most crucial learning tool, the essence of human intelligence: to forge links; to go beyond the given; to see patterns, relationship, context.
    —Marilyn Ferguson

Brainstorming

*The finest thought runs the risk of being irretrievably
forgotten if it is not written down.*
> —Arthur Schopenhauer

*Thoughts, like fleas, jump from man to man. But they
don't bite everybody.*
> —Stanislaw Lec

Writing is making meaning, a process that begins with brainstorming. Brainstorming is the most critical part of writing because it helps you generate ideas, pose questions, and gather information as you simultaneously define your subject and audience. Later, you can use brainstorming to identify gaps in your knowledge or to show where more research needs to be done.

Brainstorming is critical because it's the first time you transfer thoughts from your head onto paper. Brainstorming will help you find and identify your thoughts — not just the ones you already have, but also some new ones. Indeed, you'll hardly know many of your thoughts until you see them in writing. This is why Sartre quit writing when he lost his sight: He could no longer see his words, the symbols of his thoughts.

Brainstorming has led many scientists to brilliant ideas and major discoveries, and it can do the same for you. Here are some good ways to brainstorm:

Do **not** restrict yourself by trying to think of everything before you start. If you do, you will lose some of your thoughts because you'll be worrying about where they fit in. Similarly, do **not** write an outline and stick to it. All of these things come later. If you do them now, you'll only force yourself into thinking that you must completely understand your subject before you write about it. This ignores the ability of writing to help you understand your work. Do not try to predict what you will write.

Try **not** to formulate everything in advance. Rather, generate your ideas in any way that you please, such as with notes, lists, terms, sketches, questions, or reminders. Write in whole sentences or phrases with diagrams; this will help you discover, identify, and preserve your thoughts. This kind of brainstorming is a favorite technique of many scientists, including Charles Darwin:

> I have as much difficulty as ever in expressing myself clearly and concisely; and this difficulty has caused me a very great loss of time; but it has had the compensating advantage of forcing me to think long and intently about every sentence, and thus I have been led to see errors in reasoning and in my own observations or those of others.
>
> There seems to be a sort of fatality in my mind leading me to put at first my statement or proposition in a wrong or awkward form.

Writing with a Word Processor

> There are about as many word processing programs as there are recordings of Beethoven's Fifth—maybe more. They are all different, but some are less different than others.—Peter McWilliams, *The Word Processing Book*

Word processors have revolutionized writing by improving the speed and efficiency of writing. Although computers can't think for you, they can make it much easier and faster to type, correct, and manipulate your writing. Computers also let you bypass note cards, make quick and sweeping changes of your manuscript, change and re-examine the paper immediately, and quickly recall all parts of the paper.* Also, a computer allows you to access electronic databases such as the *Science Citation Index, Biological Abstracts,* and *Current Contents.* Writing at a computer saves time and paper, for you need no scissors, paste, ink, erasers, correction fluid, or trash cans when you write at a computer. Last-minute rewrites, once considered an impossible task, become routine.

If used properly, computers can do more than merely move text or make quick changes in your paper. They can also keep track of, format, and organize a paper, store text, justify margins, hyphenate words, build a glossary, count words and sentences, insert footnotes and headers, produce different fonts, and highlight text. Many word processing programs include spelling checkers (of up to 100,000 words), while other programs include dictionaries, graphics, grammar checkers, and a thesaurus. Style-checking programs such as *Sensible Grammar, Correct Grammar,* and *RightWriter* can check for misused words (for example, *affect* and *effect*; see Appendix 2), clichés, passive voice, and proper spacing of text. Although you don't need all of these programs to write well, you should at least know about them because they can save you

Formerly I used to think about my sentences before writing them down; but for several years I have found that it saves time to scribble in a vile hand whole pages as quickly as I possibly can, contracting half the words; and then correct deliberately. Sentences thus scribbled down are often better ones than I could have written deliberately.

Do **not** lose ideas while you struggle for the perfect phrase. If you get stumped for a specific word or phrase, mark the spot with a distinctive character (such as #). Later, use the search command of your word-processing program to locate the marker; by then, you may have discovered the right word (see "Writing with a Word-Processor," above). If you have several candidates for the right word, list them all and decide about them later. The choice will be easier after you've let some time pass.

much time and work and, in the process, free you to concentrate on your ideas rather than the drudgery of tasks such as formatting and hyphenation.

Although computers can greatly improve the efficiency of writing, they can also cause problems. For example, word processing programs allow uncontrolled cutting and pasting of identical paragraphs from one paper to another. This clutters the literature with redundancies and blocks the learning and improved communication that accompany revising and rethinking what you've written.

Computers and word-processing programs are not a substitute for knowing how to write effectively. They cannot gather information or write your paper. Similarly, style-checking programs cannot define your audience, check your logic, or meaningfully evaluate your writing. That's your job. You'll still need to read your paper carefully to ensure that all words are spelled correctly and used properly. If you don't, you're likely to include misspelled words that will make you appear either uneducated or careless. Also, remember that computer programs do not necessarily save what you write; be sure to keep a back-up copy of all your work.

Learn how to use a word processing program. However, don't write only at a computer, because, in doing so, you'll see only a small, screen-sized part of your paper at a time. Print and read a "hard copy" of your paper so you can get a broad perspective by seeing several pages at once.

*Computers allow writers to store and quickly recall huge amounts of information. Computers store information as "bytes" that represent a single character, such as a letter, symbol, or space. For example, a 40-megabyte hard-disk standard on many personal computers has enough storage to hold 300 books each 500 pages long. Another promising development is called WORM — *Write Once, Read Many* (times). This could replace a floppy disk and store as much as 700 megabytes of information. When used as digital tape, a 12-inch reel could store 1,000 gigabytes (enough to fill a billion sheets of paper) with an average access time of about 30 seconds.

Let your writing generate ideas, and think with your pencil or keyboard. Go for quantity, not quality. At this stage of writing, as in Las Vegas, never stop when you're on a roll.

Refer frequently to information you have gathered from the library, your lab notebook, reprints of relevant papers, and any other materials that might help you generate ideas.

Gather your ideas and understand your data, references, and subject. If you fail at this, you'll have nothing to say and you're doomed. Even educated, intelligent people write poorly when they have nothing to say or when they've not thought much about what they're writing.

Be sure it's the page, not your mind, that's blank. Chances are that if you can't find the right words, you haven't yet found the right idea.

Think about the topic uncritically. Write every idea you have in whatever order and form it occurs to you. Write freely and do not worry about the mechanics or finer points of grammar. Be concerned only with getting ideas from your head onto paper.

Don't think that you have to get all your ideas onto paper at one sitting. If you repeat this brainstorming process several times, you'll continue to discover new ideas and hidden angles about your subject.

Do not judge your ideas. The goal here is quantity: You'll produce at least 10 times more ideas if you write freely than if you stop and evaluate each idea. Too much self-criticism will stifle your productivity and creativity.

Remember that there is no one "right way" to write or start writing. Accept whatever comes to mind. Although an English teacher may not love what you write, don't worry about it; you'll revise it later. For now, just get the flow started.

Write answers to the questions, "What does the reader want to know about this?" and "What is important about this?" These questions should be answered in your paper because they are questions that the readers will ask as they read your paper.

Some of your ideas will come out of your head ready for battle, like Athena from the head of Zeus. However, more often your ideas will resemble weak and tottering infants needing nourishment before they can stand alone. This doesn't mean that your idea is poor. Rather, it merely indicates that you've not yet discovered all of the parts of the idea and their relationships to other ideas. Don't be too critical of yourself. Our brains can handle only a few ideas at one time. This process is analogous to learning to juggle: You can learn to juggle three or four balls, but only after much practice. Even then, you'll still be limited as to how many balls you can juggle at one time.

Many scientists try to generate ideas by making outlines. Such outlines restrict and frustrate us because they do not function the way our brain does. An outline merely lists words, while our brain works with concepts, patterns, and relationships. Unlike an outline, our brain lets ideas go in their own direction, the way branches go from the trunk of a tree. As each branch grows, we see a pattern that maps the landscape of our thoughts. Outlines are useful only *after* you start to see this landscape. Imposing an outline on your thoughts before you see relationships and patterns restricts your thoughts, thereby explaining why most people dislike outlines and why outlines usually fail to help us see the interrelationships of our ideas. But if outlines don't work, how can you discover the parts of a subject and its relationships?

One way to see the relationships of your ideas is to use *clustering*, a simple and natural way to write about and discover your thoughts. Unlike an outline, clustering imposes no structure on your thoughts. Here's how to do it:

Write a phrase summarizing the subject of your writing. Draw a circle around that phrase. I'll use clustering as an example:

Figure 2–1

The first step in clustering is to write a phrase summarizing your subject. Draw a circle around that phrase.

Then write your ideas that relate to the subject. Draw a circle around each idea and connect it with a line to related ideas. This expands our example to this:

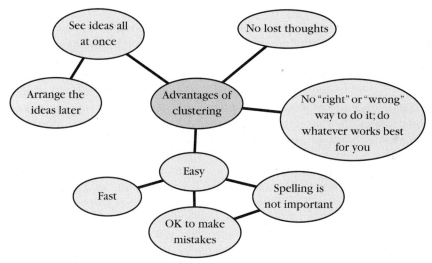

Figure 2–2

Jot down your ideas that relate to the subject. Connect each idea to related ideas.

Heed John Steinbeck's advice: "Write freely and as rapidly as possible and throw the whole thing on paper. Never correct or rewrite until the whole thing is down. Rewrite in process is usually found to be an excuse for not going on."

Brainstorm until you feel that you're repeating yourself and have thus reached a point of diminishing returns. Confront conflicts and write about everything that comes to your mind, including the obvious. Ignore the messiness and don't worry about grammar, punctuation, or style—your list of ideas is for your eyes only. Be embarrassed only when mistakes appear in the final draft.

Organizing Your Ideas

There is in writing a constant joy of sudden
discovery, of happy accident.
  —H.L. Mencken

Order and simplification are the first steps toward
the mastery of a subject. . . .
  —Thomas Mann

How do I know what I think, until I see what I say?
  —E.M. Forster

Writing is all a matter of choice and arrangement.
  —Graham Greene

The key to effective writing is organization. To write effectively, you must understand the decisions that organize your ideas and words.

Readers are seldom forced to read your papers. It's up to you to organize and write the paper so that it interests the readers. They need only enough information to know what your paper is about and why they should read it. Readers will probably study your article for its content, not for the joy of reading your prose.

The ability to efficiently organize one's ideas, and therefore one's writing, is what separates amateur writers from pros. It is also the most hidden part of the writing process. Most students are not taught organizational skills, and therefore their decisions about how to organize material often seem random or odd. This, combined with a poor understanding of other aspects of writing, causes students to waste much time struggling with organization. Experienced writers know that writing and organization are often difficult, frustrating, unpredictable, and exciting. However, they are not alarmed by their task because they understand the decision-making process used to write effectively.

The first and most basic decision you make when writing is identifying your subject and your audience. If you misjudge your subject or audience, all that follows will be extremely difficult.

Defining the Subject and Purpose of Your Writing

The discipline of the writer is to learn to be still and
listen to what his subject has to tell him.
  —Rachel Carson

Most of the knowledge and much of the genius of
the research worker lie behind his selection of what

is worth observing. It is a crucial choice, often
determining the brilliant discoverer from
the . . . plodder.
  —*Alan Gregg*

Defining the subject and purpose of your writing is the most important decision you'll make in writing, for it will influence how you write your paper. If you start in the wrong direction, you'll waste much time and effort trying to change your course. Effective writers define their subject clearly so that their readers have no trouble understanding it. In poor writing, the subject is absent, obscured, or so broad that it is unmanageable.

Brainstorming about your experiment or research will help you define the subject of many papers that you write. Other times you'll be assigned the subject for your writing assignments. Teachers usually don't make such assignments just to keep students busy. Rather, teachers make these assignments because they know that students learn when they write. When choosing your subject, remember that it is much easier to write about something that interests you than about something that bores you. You cannot write effectively about a subject if you do not understand it.

Knowing Your Audience

The first and most important step in writing is to
know who you're writing to—who they are and
what they want.
  —*David G. Lyon,* The XYZ's of Business Writing

Whom you write for doesn't change your subject, but it should affect your approach to writing. The audiences for your writing will vary greatly. For example, the audience for a term paper or laboratory report will be a professor or teaching assistant, while that for a grant proposal or publication will be biologists who know something about your research.

To identify your audience, ask yourself these questions:

Who is my audience? Whom am I writing for?

What do they already know?

What do they need to know?

When answering these questions, remember that much of what you write is for your audience's benefit and that a major goal of your writing is to enhance or expand the readers' comprehension. Therefore, start at a point from which your intended readers can follow and don't tell them what they already know. This is especially important for biologists, because biology is complex and often contains many unusual terms. These complexities increase the concentration of the writing, and therefore are amplified by poor writing. Just as archers cannot expect a target to swell to meet their poorly aimed arrows, writers cannot expect their readers to understand poorly written papers. Tiny errors made when aiming an arrow produce shots that widely miss the mark, as does sloppy writing that fails to communicate. Poor aim and complacency always cause you to miss the target, whether it be a bull's-eye or a reader's comprehension.

The responsibility for presenting understandable information rests with the writer. Therefore, aim carefully with your writing, and write to suit your audience, not yourself. Do this by using language that your readers will understand, by writing at the proper level of difficulty, and by giving readers the information they need. Start with what readers are familiar with and progress in a way that follows the readers' interests and knowledge. Knowing your audience will help you communicate better because you won't waste your readers' time by telling them what they already know.

The initial attention of your audience will be automatic if you are writing about a subject that interests them or if they must read what you have written. For example, most people who study the genetics of *Drosophila* will at least start to read a paper entitled "The Genetics of *Drosophila*." To *keep* readers' interest you must present information concisely and logically, lead readers with evidence and explanations, and not force them to reread a sentence or consult a dictionary. Such writing is both a courtesy and an insurance policy, even with a captive audience, because it ensures that your paper will be read. If you write well about an important subject, you'll have no trouble keeping readers' interest.

Gathering Information

Trouble in writing clearly . . . reflects troubled
thinking, usually an incomplete grasp of the facts or
their meaning.
    —Barbara Tuchman

Knowledge is of two kinds. We know a subject
ourselves, or we know where we can find
information upon it.
    —Samuel Johnson

Men give me credit for genius; but all the genius I
have lies in this: when I have a subject on hand, I
study it profoundly.
    —Alexander Hamilton

*If I have seen further . . . it is by standing upon the
shoulders of giants.*
        —Isaac Newton[1]

*The next best thing to knowing something is
knowing where to find it.*
        —Dr. Johnson

*Language is the only instrument of science, and
words are but the signs of ideas.*
        —Samuel Johnson

This stage of writing involves gathering data, notes, books, reprints, and other materials relevant to your subject. Gathering this information is critical to writing because, as in cooking, you can't produce a good product unless you start with the right ingredients. Fortunately, obtaining information for a biology paper is more a physical than a mental chore. All that's required is tracking down details that relate to your subject. This may involve finding references in the library, doing or repeating an experiment, documenting a method, or doing a statistical test. If you're writing about your research, much of your information will come from your lab or field notebook (see "Laboratory and Field Notebooks" p. 247).

One of the most valuable places to gather information, and one of a scientist's most valuable "laboratories," is a library. Others' ideas often clarify your thoughts, as they did for Charles Darwin:

> In October 1838 . . . I happened to read for amusement Malthus on population, and being well prepared to appreciate the struggle for existence which everywhere goes on from long continued observation of the habits of animals and plants, it at once struck me that under these circumstances favorable variations would tend to be preserved and unfavorable ones to be destroyed. The result of this would be the formation of a new species.

Unfortunately, the sheer magnitude of information in most libraries often makes it hard to find what you want. Don't be intimidated by a library: You need to know only a few things to start using it efficiently.

Card Catalog   The card catalog is a catalog of all of the library's books. These books are listed by title, author, and subject, so you need not know all of the

---

[1]Scientists often repeat this famous metaphor that Newton used to express his intellectual debt to Descartes and others who preceded him. As if to prove its own point, the metaphor was not Newton's invention. Robert Burton, an English clergyman who died before Newton was born, had written, "A dwarf standing on the shoulders of giants may see farther than a giant himself." Similarly, Burton acknowledged that 1600 years earlier the poet Lucan had written, "Pigmies placed on the shoulders of giants see more than the giants themselves."

details about a particular book to locate it. Most libraries have a computerized card catalog in which each book has a unique call number.

Many public libraries use the Dewey Decimal System, in which books are catalogued from 000 to 999 according to subject. For example, the 500 code denotes science, and 600 denotes technology and applied sciences. The 500 code is further subdivided into 570 (life sciences), 580 (botany), and 590 (zoology).

Most large libraries, including those at most universities, use the Library of Congress System to catalogue books. This system has more subdivisions and does not use long numbers such as 570.3988, as does the Dewey Decimal System. The classifications of primary interest to biologists include Q (science), R (medicine), and S (agriculture). Each of these categories is further subdivided into specific categories. For example, books about molecular biology are grouped in QH 506. Other categories of the Library of Congress System are shown in Appendix 3. You'll save yourself much time if you refer to Appendix 3 frequently while you search for information in the library.

Circulation Desk   The circulation desk is where you check books into and out of the library. This is usually the "main desk" of a library. If you want to check out a book, take the book to the circulation desk.

Reserve Desk   The reserve desk is where you can check out materials placed "on reserve" at the library. These materials typically include books, papers, and old tests. You probably cannot remove these materials from the library for more than a few hours. If your teacher tells you that he or she had something put on reserve at the library, get information about it at the reserve desk.

Reference Desk   The reference desk is where to get help in locating reference materials. Aside from journals, here are the references used most often by biologists:

Biological Abstracts is the most widely used indexing–abstracting service in the life sciences. *Biological Abstracts* publishes two issues per month and two volumes per year and adds about 300,000 citations per year from about 9,000 journals. Papers are numbered. To find the abstract, look up the paper's reference number in the accompanying volume. Papers are referenced according to author, subject, taxonomic group, and genus or genus–species names. For example, suppose that you're considering writing a paper about root gravitropism. You would start by looking up *roots* in the subject index. Figure 2–3 illustrates that section in the November 1, 1990 (Vol. 90, No. 9) issue of *Biological Abstracts*.

Now suppose you're interested in reading the paper identified by the phrase "*in not preceded by root cap as*," which is listed as reference number 104070. To find the abstract and bibliographic information for that paper, look up that reference in *Biological Abstracts* (see Figure 2–4).

104070. SACK, F. D.,* K. H. HASENSTEIN and A. BLAIR. (Dep. Botany, Ohio State Univ., Columbus, Ohio 43210.) ANN BOT (LOND) 66(2): 203–210. 1990. **Gravitropic curvature of maize roots in not preceded by root cap asymmetry.**— We tested whether the first response to gravistimulation is an asymmetry in the root tip that results from differential growth of the rootcap itself. The displacement of markers on the rootcap surface of maize (*Zea mays* L. cv. Merit) roots was quantified from videotaped images using customized software. The method was sensitive enough to detect marker displacements down to 15 μm and root curvature as early as 8 min after gravistimulation. No differential growth of the upper and lower sides of the cap occurred before or during root curvature. Fewer than a third of all gravistimulated roots developed an asymmetrical outline of the root tip after curvature had started, and this asymmetry did not occur in the rootcap itself. Our data support the view that the regions of gravitropic sensing and curvature are spatially separate during all phases of gravitropism in maize roots.

If you want to read the entire paper, look it up in *Annals of Botany*, Vol. 66, No. 2, pp. 203–210, 1990.

Citations in *Biological Abstracts* usually lag several months behind the publications. To compensate for this lag, the same publishers also provide *BioResearch Index*. Although *BioResearch Index* is not as complete or useful as *Biological Abstracts*, the listings are more up-to-date.

General Science Index includes nontechnical articles that will give you an overview of the subject and prepare you for technical articles.

Biology Digest summarizes articles selected from about 200 journals.

Index Medicus is a monthly survey of more than 2,800 journals received by the National Library of Medicine. *Index Medicus* is a major source of literature in medicine and biomedicine.

Current Contents is a weekly index that reproduces the table of contents of recent journals in many fields of biology. *Current Contents* provides a quick way to scan the titles of articles published in many journals and also lists the addresses of authors so you can write for a reprint of an article. Most of the reprint requests received by scientists can be traced to *Current Contents*.[2]

Environment Abstracts provides an excellent list of sources classified by topics.

Science Citation Index is published every two months. It lets a user look for a specific author and learn who have cited that author in their publications. Full bibliographic information about each citation is included in the *Source Index*, an accompanying volume; use the *Source Index* alone to find new papers. *Science Citation Index* is a way to identify authors studying a particular subject and is useful for tracing all references on a topic or by an author. Citations lag about a year behind publications.

Unlike many other sources, *Science Citation Index* allows a user to go forward from a particular citation. Suppose you're interested in knowing which other biologists are studying topics related to Fred Sack's work on root gravitropism. To learn this, look up SACK FD in the *Citation Index*. Figure 2–5 represents what you'll find in the November–December (No. 6C) 1990 issue. Sack's paper in *Protoplasma* was cited by D.A. Grantz in *Plant Cell*, Vol. 13, p. 667, 1990 (the "R" indicates that Grantz's paper is a review article). Similarly, Sack's paper in *American Journal of Botany* was cited by two authors in *Protoplasma*: J.Z. Kiss and

[2]Onuigbo, W.I.B. 1982. "Printer's Devil and Reprint Requests." *Journal of the American Society of Information Science* 33: 58–59.

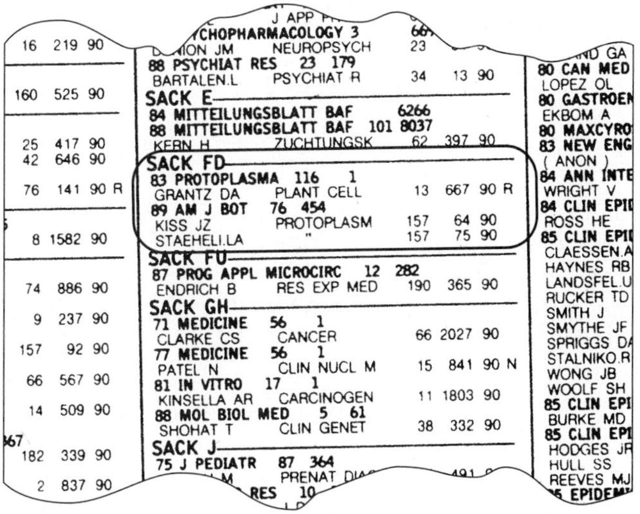
16 219 90
160 525 90
25 417 90
42 646 90
76 141 90 R
8 1582 90
74 886 90
9 237 90
157 92 90
66 567 90
14 509 90
182 339 90
2 837 90

J APP
CHOPHARMACOLOGY 3 66
ION JM NEUROPSYCH 23
88 PSYCHIAT RES 23 179
BARTALEN.L PSYCHIAT R 34 13 90
SACK E
84 MITTEILUNGSBLATT BAF 6266
88 MITTEILUNGSBLATT BAF 101 8037
KERN H ZUCHTUNGSK 62 397 90
SACK FD
83 PROTOPLASMA 116 1
GRANTZ DA PLANT CELL 13 667 90 R
89 AM J BOT 76 454
KISS JZ PROTOPLASM 157 64 90
STAEHELI.LA " 157 75 90
SACK FU
87 PROG APPL MICROCIRC 12 282
ENDRICH B RES EXP MED 190 365 90
SACK GH
71 MEDICINE 56 1
CLARKE CS CANCER 66 2027 90
77 MEDICINE 56 1
PATEL N CLIN NUCL M 15 841 90 N
81 IN VITRO 17 1
KINSELLA AR CARCINOGEN 11 1803 90
88 MOL BIOL MED 5 61
SHOHAT T CLIN GENET 38 332 90
SACK J
75 J PEDIATR 87 364
M PRENAT DIA
RES 10

80 CAN MED
LOPEZ OL
80 GASTROEN
EKBOM A
80 MAXCYRO
83 NEW ENG
(ANON)
84 ANN INTE
WRIGHT V
84 CLIN EPID
ROSS HE
85 CLIN EPID
CLAESSEN.A
HAYNES RB
LANDSFEL.U
RUCKER TD
SMITH J
SMYTHE JF
SPRIGGS DA
STALNIKO.R
WONG JB
WOOLF SH
85 CLIN EPI
BURKE MD
85 CLIN EPI
HODGES JR
HULL SS
REEVES MJ
EPIDEM

Figure 2–5
Part of a page from the November–December (No. 6C) 1990 issue of *Citation Index*. Reprinted with permission.

L.A. Staeheli. These papers are probably relevant to your interest in root gravitropism. You can get full descriptions of these articles by looking them up in the journals in which they're published or in the *Source Index* section of the *Science Citation Index*.

Other useful references for writing to learn biology include *Aquatic Sciences and Fisheries Abstracts, Chemical Abstracts, Environment Index, Biosystematic Index, Oceanic Abstracts, Pollution Abstracts, Wildlife Review,* and *Zoological Record.* If you have trouble finding reference information, ask a reference librarian for help.

Periodicals Section Included among periodicals are recent issues of journals and magazines. Although old copies of journals and magazines are usually bound and filed on shelves of the library, the most recent issues of journals are usually kept in a separate room for reading. Visit this room regularly so you can stay up-to-date on what's happening in biology.

On-Line Services Many libraries have computerized "on-line" services that can greatly speed your search for information. For example, *Bibliographic Retrieval Services* lists research papers relevant to a particular subject, while services such as *InfoTrac* list recent publications from popular journals and magazines. Similarly, BIOSIS (BioScience Information Service) Connection includes the BIOSIS Life Science Network comprising about 80 life-science databases, BIOSIS Previews, MEDLINE, Chemical Abstracts, and others, all searchable with a simple retrieval language. Other on-line services include Agricola (Agricultural Online Access) and MEDLARS (Medical Literature Analysis and Retrieval System of the National Library of Medicine; this service includes the easy-to-use *Grateful Med*). Depending on the service you use, you can request a reference list, abstracts of the articles you want, or complete copies of the articles. These searches can be done in a few minutes, and all you must provide is a list of key words. While on-line services may not identify every paper that you're interested in, they will give you a good start. A computer search usually won't identify relevant literature published before the 1960s.

To efficiently gather information in the library, you must record information systematically and completely. Here are some tips to help you efficiently gather information in the library:

Record all of the bibliographic information about the reference so that you won't have to return to the library to check on a page number or a volume of a journal. If you're using an on-line service, get a print-out of the reference. Here's an example of a complete listing of a reference (for more information about citing references, see Chapter 7):

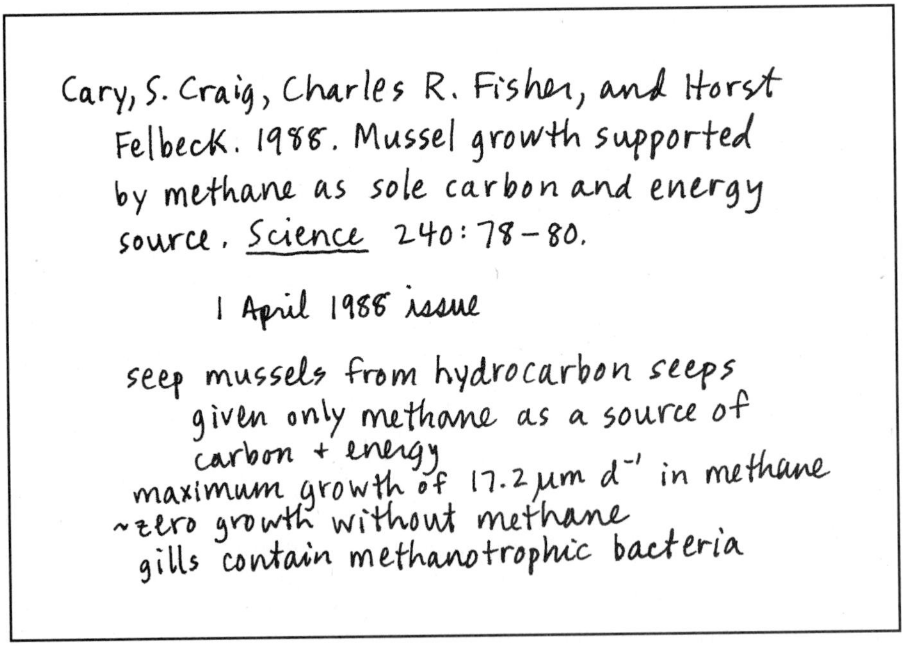

Cary, S. Craig, Charles R. Fisher, and Horst Felbeck. 1988. Mussel growth supported by methane as sole carbon and energy source. <u>Science</u> 240:78-80.

1 April 1988 issue

seep mussels from hydrocarbon seeps
given only methane as a source of
carbon + energy
maximum growth of 17.2 μm d⁻¹ in methane
~zero growth without methane
gills contain methanotrophic bacteria

Figure 2-6
Reference card containing complete bibliographic information and notes about the article.

Use index cards, and write one idea per card. You can sort these cards later. Photocopy papers that look especially useful. If you prefer to take notes on full-size pages, begin each topic on a separate page. On each card write the usefulness of and important information contained in the reference. These cards, containing bibliographic details and information about the article or book, comprise an annotated bibliography. Such a bibliography will help you record and coordinate resources and information.

If a book or paper is unavailable in your library, ask the reference librarian if he or she can help you get the book or paper via an interlibrary loan. Note on the index card where you obtained the book or paper.

Unless you want to quote a paper or book, take notes in your own words. This will help you to get away from being awed by other people's words and to become more comfortable with your own ideas and writing. This, along with taking notes in incomplete sentences, will also help you to avoid accidental plagiarism.[3] Moreover, it will force you to understand the material. If you don't understand the material, you will be unable to take good notes in your own words.

Make sure your notes are complete so that you know the context of the author's ideas. Don't use too much shorthand: You might forget what your abbreviations mean.

Nothing discredits a writer more than a mistake. When readers discover such mistakes, they conclude that all of the writer's work may be riddled with mistakes. Fortunately, accurate writing requires more effort than brilliance. Check the accuracy of all your facts.

If you gather information from scientific journals, you'll realize that a relatively small number of journals contain much of the important information about your topic. For example, papers of the 300 most-cited biologists for 1961–76 appeared in only 86 journals. Five of these journals accounted for 10% of the citations, and 10 of the journals accounted for half of the citations.[4] Similarly, in a bibliography on schistosomiasis (snail fever, caused by an infection by blood flukes) covering 110 years and containing 10,286 references, almost half of the references were from only 50 of the 1,738 journals cited.[5]

Although gathering information is relatively easy, don't underestimate its

---

[3]Much plagiarism, rather than being deliberate, results from poor notes, wishful thinking, or false recollections. However, accidental plagiarism *is* plagiarism and cannot be accepted any more than can a local, unknown rock-band claiming that it wrote "Stairway to Heaven" or "Layla." Plagiarism is equivalent to stealing. Report your observations honestly and in your own words.

[4]Garfield, E. 1977–1983. "The 300 Most Cited Authors, 1961–1976, Including Co-authors. 3A. Their Most Cited Papers. Introduction and Journal Analysis." In: *Essays of an Information Scientist.* Philadelphia: ISI Press, 3:689–700.

[5]Warren, K.S. 1981. "Selective Aspects of the Biomedical Literature." In: *Coping with the Biomedical Literature.* New York: Praeger.

importance. Nothing wastes more of a writer's time than the paralysis that accompanies not having details to document what you want to say.

Starting to Write

Writing is easy. All you have to do is stare at a
blank sheet of paper until drops of blood form on
your forehead.
—Gene Fowler

He has half the deed done who has made a beginning
—Horace

No matter how much you've gathered information, organized your thoughts, or researched a topic, there comes a time when you must face a blank page or computer screen. Many people have much trouble starting a writing assignment. As soon as they sit down to write, they instead find themselves in front of a vending machine, walking down the hall, or deciding that they can't start writing until they rearrange their papers, wash their car, or sharpen all of their pencils. This problem is referred to as "writer's block" and usually results from writers either not knowing what they're writing about or trying to get everything onto paper at one sitting. No one, not even the world's best writers, can write things in final form at the first attempt. Rather, good writers write effectively by using a process involving organizing their ideas, creating a tentative outline, producing a first draft, and, finally, revising their work.

Organizing Your Ideas

A useful aid in getting a clear understanding of a
problem is to write a report on all the information
available. This is helpful when one is starting an
investigation, when up against a difficulty, or when
the investigation is nearing completion. Also at the
beginning of an investigation it is useful to set out
clearly the questions for which an answer is being
sought. Stating the problem precisely sometimes
takes one a long way toward the solution. The
systematic arrangement of the data often discloses
flaws in the reasoning, or alternative lines of thought
which had been missed. Assumptions and
conclusions at first accepted as "obvious" may even

prove indefensible when set down clearly and
examined critically.
  —W.I.B. Beveridge, *The Art of Scientific
    Investigation*

I spend a good deal of time over the general
arrangement of the matter. I first make the rudest
outline in two or three pages, and then a larger one
in several pages, a few words or one word standing
for a whole discussion or series of facts. Each one of
these headings is again enlarged and often
transferred before I begin to write in extenso. As in
several of my books facts observed by others have
been very extensively used, and as I have always
had several quite distinct subjects in hand at the
same time, I may mention that I keep from thirty to
forty large portfolios, in cabinets with labeled shelves,
into which I can at once put a detached reference or
memorandum. I have bought many books, and at
their ends I make an index of all the facts that
concern my work; or, if the book is not my own,
write out a separate abstract, and of such abstracts I
have a large drawer full. Before beginning on any
subject I look to all the short indexes and make a
general and classified index, and by taking the one
or more proper portfolios I have all the information
collected during my life ready for use.
  —Charles Darwin

The card-player begins by arranging his hand for
maximum sense. Scientists do the same thing with
the facts they gather.
  —Isaac Asimov

Science is organized knowledge, and writing is the means to that organization. Organizing your ideas is an exercise in labeling your ideas and reviewing
the products of your brainstorming. Begin by reading your notes. Many of your
ideas will be irrelevant to your subject or audience and should be eliminated, no
matter how interesting. Also eliminate ideas that are relevant but insignificant. If
you include those ideas, you'll imply that all of your ideas have the same
importance. In doing this you will lose your readers before you can spring the
punch line.
  Try to discover a natural, commonsense sequence among the topics. You
may find your ideas easier to shuffle and rearrange if they are listed on index
cards, one idea or example per card. You can then divide and subdivide your

subject by moving the cards around until they fall into a logical pattern. Although this kind of "writing architecture" seems simple, it is critical because it commits you to a specific course of action. You can later overcome mistakes in execution, but you can't afford mistakes in architecture. If you fail to label and organize your thoughts, you won't be satisfied with your paper no matter how well you build it. Similarly, if you haven't taken the time to organize your thoughts, you'll probably never find time to revise your paper once it's written. But that's exactly what you'll have to do if you string ideas together randomly or without forethought.

Effectively labeling your thoughts also simplifies the next step of the writing process: creating a tentative outline of your paper.

Creating a Tentative Outline

The first rule of style is to have something to say.
The second rule of style is to control yourself when,
by chance, you have two things to say; say first
one, then the other, not both at the same time.
—George Polya

Labeling your thoughts produces parts of an outline that almost puts itself together. This is much easier than trying to create an outline as the first step in writing, which often limits your perspective. When making your outline, don't be handcuffed by a formal structure of roman numerals or numbered subtopics. Focus instead on the *function* of the outline, which is to help you organize and manipulate subjects so you can identify your ideas and communicate clearly.

Preparing an outline is a powerful way to decide what elements are essential to your idea and to fix their logical relationships. It is also an efficient way to determine the key points of your writing. As John Jordan has cogently put it:

An outline is in part a pious statement of intention, in part a skeleton to flesh out, in part a blueprint to follow, in part a cut-down model on which to try out ideas. It is not, except under unusual circumstances, a deep freeze to preserve bits of prime verbiage for later embellishment of the finished work. If you think of an outline as a skeleton upon which to build up a structure—like a sculptor's armature for supporting and directing the shape of a clay statue—you might remember that a skeleton does not show through except in case of emaciation. Yet bone structure is of fundamental importance, none the less. If you think of an outline as an architect's drawing, you might remember that even architects make mistakes and very few buildings are put up without some changes in plan during construction. An outline is not a strait jacket, it is a tool; its only real requirement is that it be useful.

Your outline is a skeleton of the first draft of your paper; it will tell you what you've done and suggest the outcome of your paper. It is simple to create: It's merely the logical listing of the ideas that you've already labeled. Here's how to do it:

1. Divide your cards into categories that serve as building blocks. For example, in a paper about tropisms in plants, such categories could include the following subjects:

   Phototropism

   Gravitropism

   Hydrotropism

2. Arrange the piles of cards into a logical sequence. In a discussion of gravitropism, the sequence might be arranged as follows:

   How plants perceive gravity

   How plants transform gravity into a physiological signal

   How that physiological signal induces gravitropism

3. Arrange the cards in each pile into a logical sequence. The pile of cards in the stack labeled "How the physiological signal induces gravitropism" might be arranged as follows:

   The identity of the signal

   Where the signal originates

   Where the signal exerts its effect

   Other factors that influence the effect of the signal

This forms a tentative outline for your paper. Stripped to its essentials, the outline should highlight the logic of your ideas. Ask yourself these questions to determine if your outline is in good shape:

Are the topics balanced? Have you devoted too much or too little space to any topics?

Have you abruptly jumped from one subject to another?

Are there gaps in the logic? If the gaps are Grand Canyons, rethink your plan.

Did you gather enough information?

Have you discovered all of your ideas?

Continue to brainstorm as you examine your outline. Is your outline complete? If not, add other topics. Similarly, delete all irrelevant topics. Your outline will tell you if you're on the right track. If you're on the wrong track, the outline will tell you that also. Your outline will also help you divide your paper into sections and subsections that alert the reader to what's coming.

Take the time now to rethink, clarify, and reorganize your ideas; it will pay dividends later.

Going from Design to Construction: Creating a First Draft

*Science consists in grouping facts so that general
laws or conclusions may be drawn from them.*
    —Charles Darwin

You're now ready to piece together a first draft of your paper. To do this, all you need to do is expand and build up each section of the outline, one section at a time. Again, don't worry about grammar, details, or style. Parts will be vague, while others will seem disjointed and thin. Accept that the first draft of your paper will need more work: Just get your ideas into a logically arranged text. At this stage of writing, as in previous ones, writing continues to be a means of thinking, thereby helping you to create more knowledge than you've been given.

If you have trouble writing the first draft of your paper, don't accept the self-delusion of excuses such as, "I know this but just can't find the right words." Instead, carefully review what you've done. Have you gathered enough information? Have you organized the information logically? Don't accept the excuse of "writer's block": If you've gathered the information or completed your experiments, your ideas are already there. Any trouble you have in writing about your ideas probably results from "science block" rather than from "writer's block."

Reviewing Your First Draft

Read your first draft and give yourself a mental pat on the back. Although what you've written looks rough and still needs work, you've accomplished the most critical part of the work. You're halfway home.

As you reread your paper you'll probably be tempted to add to your notes, thus beginning the next stage of the writing process: revising. However, for now just concentrate on the big picture and make sure that you've gotten the main idea, content, and organization. Then ask yourself these questions:

Do you really have something to say? This question, which will focus your thinking, is what your readers will ask as they read your paper. No one will care how well or badly you express yourself if you have nothing to say. Moreover, it is impossible to write well about anything if you have nothing to say.

Clear writing requires clear thinking. However, if your ideas are a bit muddled when you start to write, the very act of writing may help clarify your thinking and thereby clarify your writing. Most of your ideas will not become clear until you've written them down.

Have you told readers everything they need to know? It's easy for writers, when immersed in a subject, to lose sight of how much background the reader needs to grasp their ideas. Too much detail bores readers, while too little confuses them. Be sure that you've written your paper for your audience.

Did you stray from your outline or make changes that detract from an orderly presentation? If so, ask yourself why you made those changes. They'll probably need to be corrected, but they may indicate new ideas that should stay in the paper.

If you're writing a paper for publication, ask yourself, "Have I answered an important biological question?" We make little progress with "potboiler" publications containing randomly gathered facts that writers wishfully claim may "shed some light" on a nonexistent problem.

If the answers to these questions are no, rethink your plans. No amount of writing can disguise poorly designed experiments, an author having nothing to say, or an author who doesn't understand what he or she is writing about. For example, here's what biology students wrote when they tried to write about a subject they didn't quite understand:

A fossil is an extinct animal. The older it is, the more extinct it is.

Artificial insemination is when the farmer does it to the cow and not the bull.

When you breathe, you inspire. When you do not breathe, you expire.

Think about what you've written and check for simple, but subtle, mistakes. For example, one student wrote a note to me stating, "I have a temperature." I wrote her a reply saying that I, too, had a temperature, as did everything else in the universe. Similarly, another biologist wrote, "The solvents were evaporated *in vacuo* at 40°C under a stream of nitrogen." Although including the phrase *in vacuo* makes this sentence sound sophisticated, it also helps hide the absurdity of a vacuum made of nitrogen. Finally, a safety-conscious student wrote in his laboratory report, "Goggles are required to do this experiment." I wonder what grade those goggles made on the experiment.

If you have told readers what they need to know, if you have presented information in a logical way, and if you have answered a significant biological question, then you're ready for the next stage of the writing process: revising and rethinking what you've written so that you'll communicate effectively with your readers.

Stephen Jay Gould

False Premise, Good Science

Stephen Jay Gould's (b. 1941) award-winning
essays and books describe riddles of evolution. His
essay "False Premise, Good Science" from *The
Flamingo's Smile: Reflections on Natural History* is
reprinted here.

My vote for the most arrogant of all scientific titles goes without hesitation to a famous paper written in 1866 by Lord Kelvin, "The 'Doctrine of Uniformity' in Geology Briefly Reputed." In it, Britain's greatest physicist claimed that he had destroyed the foundation of an entire profession not his own. Kelvin wrote:

> The "Doctrine of Uniformity" in Geology, as held by many of the most eminent of British geologists, assumes that the earth's surface and upper crust have been nearly as they are at present in temperature and other physical qualities during millions of millions of years. But the heat which we know, by observation, to be now conducted out of the earth yearly is so great, that if *this* action had been going on with any approach to uniformity for 20,000 million years, the amount of heat lost out of the earth would have been about as much as would heat, by 100° Cent., a quantity of ordinary surface rock of 100 times the earth's bulk. (See calculation appended.) This would be more than enough to melt a mass of surface rock equal in bulk to the *whole earth*. No hypothesis as to chemical action, internal fluidity, effects of pressure at great depth, or possible character of substances in the interior of the earth, possessing the smallest vestige of probability, can justify the supposition that the earth's crust has remained nearly as it is, while from the whole, or from any part, of the earth, so great a quantity of heat has been lost.

I apologize for inflicting so long a quote so early in the essay, but this is not an extract from Kelvin's paper. It is the whole thing (minus the appended calculation). In a mere paragraph, Kelvin felt he had thoroughly undermined the very basis of his sister discipline.

Kelvin's arrogance was so extreme, and his later comeuppance so spectacular, that the tale of his 1866 paper, and of his entire, relentless forty-year campaign for a young earth, has become the classical moral homily of our geological textbooks. But beware of conventional moral homilies. Their probability of accuracy is about equal to the chance that George Washington really scaled that silver dollar clear across the Rappahannock.

The story, as usually told, goes something like this. Geology, for several centuries, had languished under the thrall of Archbishop Ussher and his biblical chronology of but a few thousand years for the earth's age. This restriction of time led to the unscientific doctrine of catastrophism—the notion that miraculous upheavals and paroxysms must characterize our earth's history if its entire geological story must be compressed into the Mosaic chronology. After long struggle, Hutton and Lyell won the day for science with their alternative idea of uniformitarianism—the claim that current rates of change, extrapolated over limitless time, can explain all our history from a scientific standpoint by direct observation of present processes and their results. Uniformity, so the story goes, rests on two propositions: essentially unlimited time (so that slow processes can achieve their accumulated effect), and an earth that does not alter its basic form and style of change throughout this vast time. Uniformity in geology led to evolution in biology and the scientific revolution spread. If we deny uniformity, the homily continues, we undermine science itself and plunge geology back into its own dark ages.

Yet Kelvin, perhaps unaware, attempted to undo this triumph of scientific geology. Arguing that the earth began as a molten body, and basing his calculation upon loss of heat from the earth's interior (as measured, for example, in mines), Kelvin recognized that the earth's solid surface could not be very old—probably 100 million years, and 400 million at most (although he later revised the estimate downward, possibly to only 20 million years). With so little time to harbor all of evolution—not to mention the physical history of solid rocks—what recourse did geology have except to its discredited idea of catastrophes? Kelvin had plunged geology into an inextricable dilemma while clothing it with all the prestige of quantitative physics, queen of the sciences. One popular geological textbook writes (C. W. Barnes), for example:

> Geologic time, freed from the constraints of literal biblical interpretation, had become unlimited; the concepts of uniform change first suggested by Hutton now embraced the concept of the origin and evolution of life. Kelvin single-handedly destroyed, for a time, uniformitarian and evolutionary thought. Geologic time was still restricted because the laws of physics bound as tightly as biblical literalism ever had.

Fortunately for a scientific geology, Kelvin's argument rested on a false premise—the assumption that the earth's current heat is a residue of its original molten state and not a quantity constantly renewed. For if the earth continues to generate heat, then the current rate of loss cannot be used to infer an ancient condition. In fact, unbeknown to Kelvin, most of the earth's internal heat is newly generated by the process of radioactive decay. However elegant his calculations, they were based on a false premise, and Kelvin's argument collapsed with the discovery of radioactivity early in our century. Geologists should have trusted their own intuitions from the start and not bowed before the false lure of physics. In any case, uniformity finally won and scientific geology was restored. This transient episode teaches us that we must trust the careful empirical data of a profession and not rely too heavily on theoretical interventions from outside, whatever their apparent credentials.

So much for the heroic mythology. The actual story is by no means so simple or as easily given an evident moral interpretation. First of all, Kelvin's arguments, although fatally flawed as outlined above, were neither so coarse nor as unacceptable to geologists as the usual story goes. Most geologists were inclined to treat them as a genuine reform of their profession until Kelvin got carried away with further restrictions upon his original estimate of 100 million years. Darwin's strong opposition was a personal campaign based on his own extreme gradualism, not a consensus. Both Wallace and Huxley accepted Kelvin's age and pronounced it consonant with evolution. Secondly, Kelvin's reform did not plunge geology into an unscientific past, but presented instead a different *scientific* account based on another concept of history that may be more valid than the strict uniformitarianism preached by Lyell. Uniformitarianism, as advocated by Lyell, was a specific and restrictive theory of history, not (as often misunderstood) a general account of how science must operate. Kelvin had attacked a legitimate target.

Kelvin's Arguments and the Reaction of Geologists

As codiscoverer of the second law of thermodynamics, Lord Kelvin based his arguments for the earth's minimum age on the dissipation of the solar system's original energy as heat. He advanced three distinct claims and tried to form a single quantitative estimate for the earth's age by seeking agreement among them (see Joe D. Burchfield's *Lord Kelvin and the Age of the Earth*, the source for most of the technical information reported here).

Kelvin based his first argument on the age of the sun. He imagined that the sun had formed through the falling together of smaller meteoric masses. As these meteors were drawn together by their mutual gravitational attraction, their potential energy was transformed into kinetic energy, which, upon collision, was finally converted into heat, causing the sun to shine. Kelvin felt that he could calculate the total potential energy in a mass of meteors equal to the sun's bulk and, from this, obtain an estimate of the sun's original heat. From this estimate, he could calculate a minimum age for the sun, assuming that it has been shining at its present intensity since the beginning. But this calculation was crucially dependent on a set of factors that Kelvin could not really estimate—including the original number of meteors and their original distance from each other—and he never ventured a precise figure for the sun's age. He settled on a number between 100 and 500 million years as a best estimate, probably closer to the younger age.

Kelvin based his second argument on the probable age of the earth's solid crust. He assumed that the earth had cooled from an originally molten state and that the heat now issuing from its mines recorded the same process of cooling that had caused the crust to solidify. If he could measure the rate of heat loss from the earth's interior, he could reason back to a time when the earth must have contained enough heat to keep its globe entirely molten—assuming that this rate of dissipation had not changed through time. (This is the argument for his "brief" refutation of uniformity, cited at the beginning of this essay.) This argument sounds more "solid" than the first claim based on a hypothesis about how the sun formed. At least one can hope to measure directly its primary ingredient—the earth's current loss of heat. But Kelvin's second argument still depends upon several crucial and unprovable assumptions about the earth's composition. To make his calculation work, Kelvin had to treat the earth as a body of virtually uniform composition that had solidified from the center outward and had been, at the time its crust formed, a solid sphere of similar temperature throughout. These restrictions also prevented Kelvin from assigning a definite age for the solidification of the earth's crust. He ventured between 100 and 400 million years, again with a stated preference for the smaller figure.

Kelvin based his third argument on the earth's shape as a spheroid flattened at the poles. He felt that he could relate this degree of polar shortening to the speed of the earth's rotation when it formed in a molten state amenable to flattening. Now we know—and Kelvin knew also—that the earth's rotation has been slowing down continually as a result of tidal friction. The earth rotated more rapidly when it first formed. Its current shape should therefore indicate its age. If the earth formed a long time ago, when rotation was quite rapid, it should now be very flat. If the earth is not so ancient, then it formed at a rate of rotation not so different from its current pace, and flattening should be less. Kelvin felt that the small degree of actual flattening indicated a relatively young age for the earth. Again, and for the third time, Kelvin based his argument upon so many unprovable assumptions (about the earth's uniform composition, for example) that he could not calculate a precise figure for the earth's age.

Thus, although all three arguments had a quantitative patina, none was precise. All depended upon simplifying assumptions that Kelvin could not justify. All therefore yielded only vague estimates with large margins of error. During most of Kelvin's forty-year campaign, he usually cited a figure of 100 million years for the earth's age—plenty of time, as it turned out, to satisfy nearly all geologists and biologists.

Darwin's strenuous opposition to Kelvin is well recorded, and later commentators have assumed that he spoke for a troubled consensus. In fact, Darwin's antipathy to Kelvin was idiosyncratic and based on the strong personal commitment to gradualism so characteristic of his world view. So wedded was Darwin to the virtual necessity of unlimited time as a prerequisite for evolution by natural selection that he invited readers to abandon *The Origin of Species* if they could not accept this premise: "He who can read Sir Charles Lyell's grand work on the *Principles of Geology*, and yet does not admit how incomprehensively vast have been the past periods of time, may at once close this volume." Here Darwin commits a fallacy of reasoning—the confusion of gradualism with natural selection—that characterized all his work and that inspired Huxley's major criticism of the *Origin*: "You load yourself with an unnecessary difficulty in adopting *Natura non facit saltum* [Nature does not proceed by leaps] so unreservedly." Still, Darwin cannot be entirely blamed, for Kelvin made the same error in arguing explicitly that his young age for the earth cast grave doubt upon natural selection as an evolutionary mechanism (while not arguing against evolution itself). Kelvin wrote:

> The limitations of geological periods, imposed by physical science, cannot, of course, disprove the hypothesis of transmutation of species; but it does seem sufficient to disprove the doctrine that transmutation has taken place through "descent with modification by natural selection."

Thus, Darwin continued to regard Kelvin's calculation of the earth's age as perhaps the gravest objection to his theory. He wrote to Wallace in 1869 that "Thomson's [Lord Kelvin's] views on the recent age of the world have been for some time one of my sorest troubles." And, in 1871, in striking metaphor, "But then comes Sir W. Thomson like an odious spectre." Although Darwin generally stuck to his guns and felt in his heart of hearts that something must be wrong with Kelvin's calculations, he did finally compromise in the last edition of the *Origin* (1872), writing that more rapid changes on the early earth would have accelerated the pace of evolution, perhaps permitting all the changes we observe in Kelvin's limited time:

It is, however, probable, as Sir William Thompson [*sic*] insists, that the world at a very early period was subjected to more rapid and violent changes in its physical conditions than those now occurring; and such changes would have tended to induce changes at a corresponding rate in the organisms which then existed.

Darwin's distress was not shared by his two leading supporters in England, Wallace and Huxley. Wallace did not tie the action of natural selection to Darwin's glacially slow time scale; he simply argued that, if Kelvin limited the earth to 100 million years, then natural selection must operate at generally higher rates than we had previously imagined. "It is within that time [Kelvin's 100 million years], therefore, that the whole series of geological changes, the origin and development of all forms of life, must be compressed." In 1870, Wallace even proclaimed his happiness with a time scale of but 24 million years since the inception of our fossil record in the Cambrian explosion.

Huxley was even less troubled, especially since he had long argued that evolution might occur by saltation, as well as by slow natural selection. Huxley maintained that our conviction about the slothfulness of evolutionary change had been based on false and circular logic in the first place. We have no independent evidence for regarding evolution as slow; this impression was only an inference based on the assumed vast duration of fossil strata. If Kelvin now tells us that these strata were deposited in far less time, then our estimate of evolutionary rate must be revised correspondingly.

> Biology takes her time from geology. The only reason we have for believing in the slow rate of the change in living forms is the fact that they persist through a series of deposits which, geology informs us, have taken a long while to make. If the geological clock is wrong, all the naturalist will have to do is to modify his notions of the rapidity of change accordingly.

Britain's leading geologists tended to follow Wallace and Huxley rather than Darwin. They stated that Kelvin had performed a service for geology in challenging the virtual eternity of Lyell's world and in "restraining the reckless drafts" that geologists so rashly make on the "bank of time," in T. C. Chamberlin's apt metaphor. Only late in his campaign, when Kelvin began to restrict his estimate from a vague and comfortable 100 million years (or perhaps a good deal more) to a more rigidly circumscribed 20 million years or so did geologists finally rebel. A. Geikie, who had been a staunch supporter of Kelvin, then wrote:

> Geologists have not been slow to admit that they were in error in assuming that they had an eternity of past time for the evolution of the earth's history. They have frankly acknowledged the validity of the physical arguments which go to place more or less definite

limits to the antiquity of the earth. They were, on the whole, disposed to acquiesce in the allowance of 100 millions of years granted them by Lord Kelvin, for the transaction of the long cycles of geological history. But the physicists have been insatiable and inexorable. As remorseless as Lear's daughters, they have cut down their grant of years by successive slices, until some of them have brought the number to something less than ten millions. In vain have geologists protested that there must be somewhere a flaw in a line of argument which tends to results so entirely at variance with the strong evidence for a higher antiquity.

Kelvin's Scientific Challenge and the Multiple Meanings of Uniformity

As a master of rhetoric, Charles Lyell did charge that anyone who challenged his uniformity might herald a reaction that would send geology back to its prescientific age of catastrophes. One meaning of uniformity did uphold the integrity of science in this sense—the claim that nature's laws are constant in space and time, and that miraculous intervention to suspend these laws cannot be permitted as an agent of geological change. But uniformity, in this methodological meaning, was no longer an issue in Kelvin's time, or even (at least in scientific circles) when Lyell first published his *Principles of Geology* in 1830. The scientific catastrophists were not miracle mongers, but men who fully accepted the uniformity of natural law and sought to render earth history as a tale of *natural* calamities occurring infrequently on an ancient earth.

But uniformity also had a more restricted, substantive meaning for Lyell. He also used the term for a particular theory of earth history based on two questionable postulates: first, that rates of change did not vary much throughout time and that slow and current processes could therefore account for all geological phenomena in their accumulated impact; second, that the earth had always been about the same, and that its history had no direction, but represented a steady state of dynamically constant conditions.

Lyell, probably unconsciously, then performed a clever and invalid trick of argument. Uniformity had two distinct meanings—a methodological postulate about uniform laws, which all scientists had to accept in order to practice their profession, and a substantive claim of dubious validity about the actual history of the earth. By calling them both uniformity, and by showing that all scientists were uniformitarians in the first sense, Lyell also cleverly implied that, to be a scientist, one had to accept uniformity in its substantive meaning as well. Thus, the myth developed that any opposition to uniformity could only be a rearguard action against science itself—and the impression arose that if Kelvin was attacking the "doctrine of uniformity" in geology, he must represent the forces of reaction.

In fact, Kelvin fully accepted the uniformity of law and even based his calculations about heat loss upon it. He directed his attack against uniformity only upon the substantive (and dubious) side of Lyell's vision. Kelvin advanced two complaints about this substantive meaning of uniformity. First, on the question of rates. If the earth were substantially younger than Lyell and the strict uniformitarians believed, then modern, slow rates of change would not be sufficient to render its history. Early in its history, when the earth was hotter, causes must have been more energetic and intense. (This is the "compromise" position that Darwin finally adopted to explain faster rates of change early in the history of life.) Second, on the question of direction. If the earth began as a molten sphere and lost heat continually through time, then its history had a definite pattern and path of change. The earth had not been perennially the same, merely changing the position of its lands and seas in a never-ending dance leading nowhere. Its history followed a definite road, from a hot, energetic sphere to a cold, listless world that, eventually, would sustain life no longer. Kelvin fought, within a scientific context, for a *short-term, directional* history against Lyell's vision of an essentially eternal steady-state. Our current view represents the triumph of neither vision, but a creative synthesis of both. Kelvin was both as right and as wrong as Lyell.

Radioactivity and Kelvin's Downfall

Kelvin was surely correct in labeling as extreme Lyell's vision of an earth in steady-state, going nowhere over untold ages. Yet, our modern time scale stands closer to Lyell's concept of no appreciable limit than to Kelvin's 100 million years and its consequent constraint on rates of change. The earth is 4.5 billion years old.

Lyell won this round of a complicated battle because Kelvin's argument contained a fatal flaw. In this respect, the story as conventionally told has validity. Kelvin's argument was not an inevitable and mathematically necessary set of claims. It rested upon a crucial and untested assumption that underlay all Kelvin's calculations. Kelvin's figures for heat loss could measure the earth's age only if that heat represented an original quantity gradually dissipated through time—a clock ticking at a steady rate from its initial reservoir until its final exhaustion. But suppose that new heat is constantly created and that its current radiation from the earth reflects no original quantity, but a modern process of generation. Heat then ceases to be a gauge of age.

Kelvin recognized the contingent nature of his calculations, but the physics of his day included no force capable of generating new heat, and he therefore felt secure in his assumption. Early in his campaign, in calculating the sun's age, he had admitted his crucial dependence upon no new source of energy, for he had declared his results valid "unless new sources now unknown to us are prepared in the great storehouse of creation."

Then, in 1903, Pierre Curie announced that radium salts constantly release newly generated heat. The unknown source had been discovered. Early students of radioactivity quickly recognized that most of the earth's heat must be continually generated by radioactive decay, not merely dissipating from an originally molten condition — and they realized that Kelvin's argument had collapsed. In 1904, Ernest Rutherford gave this account of a lecture, given in Lord Kelvin's presence, and heralding the downfall of Kelvin's forty-year campaign for a young earth:

> I came into the room, which was half dark, and presently spotted Lord Kelvin in the audience and realized that I was in for trouble at the last part of the speech dealing with the age of the earth, where my views conflicted with his. To my relief, Kelvin fell fast asleep, but as I came to the important point, I saw the old bird sit up, open an eye and cock a baleful glance at me! Then a sudden inspiration came, and I said Lord Kelvin had limited the age of the earth, provided no new source of heat was discovered. That prophetic utterance refers to what we are now considering tonight, radium!

Thus, Kelvin lived into the new age of radioactivity. He never admitted his error or published any retraction, but he privately conceded that the discovery of radium had invalidated some of his assumptions.

The discovery of radioactivity highlights a delicious double irony. Not only did radioactivity supply a new source of heat that destroyed Kelvin's argument; it also provided the clock that could then measure the earth's age and proclaim it ancient after all! For radioactive atoms decay at a constant rate, and their dissipation does measure the duration of time. Less than ten years after the discovery of radium's newly generated heat, the first calculations for radioactive decay were already giving ages in billions of years for some of the earth's oldest rocks.

We sometimes suppose that the history of science is a simple story of progress, proceeding inexorably by objective accumulation of better and better data. Such a view underlies the moral homilies that build our usual account of the advance of science — for Kelvin, in this context, clearly impeded progress with a false assumption. We should not be beguiled by such comforting and inadequate stories. Kelvin proceeded by using the best science of his day, and colleagues accepted his calculations. We cannot blame him for not knowing that a new source of heat would be discovered. The framework of his time included no such force. Just as Maupertuis lacked a proper metaphor for recognizing that embryos might contain coded instructions rather than preformed parts, Kelvin's physics contained no context for a new source of heat.

The progress of science requires more than new data; it needs novel frameworks and contexts. And where do these fundamentally new views of the world arise? They are not simply discovered by pure observation; they require new modes of thought. And where can we find them, if old modes do not even include the right metaphors? The nature of true genius must lie in the elusive capacity to construct these new modes from apparent darkness. The basic chanciness and unpredictability of science must also reside in the inherent difficulty of such a task.

What stylistic techniques does Gould use to engage his audience? What is "progress" in science? Can false premise lead to good science? Why or why not?

Exercises

1. Use clustering and brainstorming to gather your ideas about the following quotations about science and biology:

> Destroying species is like tearing pages out of an unread book, written in a language humans hardly know how to read, about the place where they live. — Robert Holmes III

> [Biology] is the least self-centered, the least narcissistic of the sciences — the one that, by taking us out of ourselves, leads us to re-establish the link with nature and to shake ourselves free from our spiritual isolation. — Jean Rostand

> The role of biology today, like the role of every other science, is simply to describe, and when it explains it does not mean that it arrives at finality; it only means that some descriptions are so charged with significance that they expose the relationship of cause and effect. — Donald Culross Peattie

> We prefer economic growth to clean air. — Charles Barden

> Life-cycle completion is indeed the master law governing all the activities of the organism, to which other laws of smaller scope, such as the law of need-satisfaction, are subordinate. — E.S. Russell

> I . . . object to dividing the study of living processes into botany, zoology, and microbiology because by any such arrangement, the interrelations within the biological community get lost. Corals cannot be studied without reference to the algae that live with them; flowering plants without the insects that pollinate them; grasslands without the grazing animals. — Marston Bates

> Society increasingly has neglected the substructure of biology, to its own peril. — Edward O. Wilson

> Teleology is a lady without whom no biologist can live; yet he is ashamed to show himself in public with her. — Franklin von Bruecke

2. How are *Current Contents* and *Biological Abstracts* similar? How are they different? Which do you think is most useful and why?

3. Go to the library and examine an issue of *Current Contents*. How could you use *Current Contents* to write and learn about biology?

4. Locate *Biological Abstracts* in the library. Use the subject index to locate five articles about a particular biological topic. Make a reference card for each of the articles.

5. Choose any topic of biology and define it for an audience of your classmates. Write the following about the subject:

    Five statements

    Three yes–no questions

    A thesis statement based on one of these questions

    Two facts supporting this statement

    A paragraph summarizing the subject

6. Choose a subject, create an audience, and brainstorm for 20 minutes. Then organize your ideas and create a tentative outline for the subject.

7. Should biologists be allowed to do research that involves killing animals? Write a short essay describing both sides of this controversial issue.

8. The human population is growing extremely fast. Consider these data:

Year	Population (millions)
8000 B.C.	5
400 B.C.	86
1 A.D.	133
1650	545
1750	728
1800	906
1850	1130
1900	1610
1950	2400
1960	2998
1970	3659
1980	4551
1990	5300
2000	6500+ (projected)

Write a short essay describing the importance of these data. Do not repeat the data; rather, discuss what they mean.

9. Write a short essay describing how this *The Far Side* cartoon relates to biology:

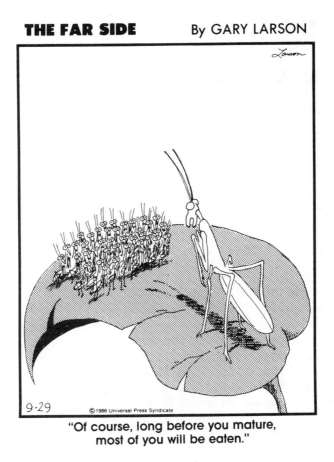

"Of course, long before you mature, most of you will be eaten."

Figure 2–7
The Far Side. Copyright 1986. UNIVERSAL PRESS SYNDICATE. Reprinted with permission.

CHAPTER THREE

Revising Is Rethinking:

Writing to Communicate

. . . [T]o eliminate the vice of wordiness is to
ensure the virtue of emphasis, which depends more
on conciseness than on any other factor. Wherever
we can make twenty-five words do the work of
fifty, we halve the area in which looseness and
disorganization can flourish, and by reducing the
span of attention required we can increase the
force of thought. To make our words count for as
much as possible is surely the simplest as well as
the hardest secret of style.
　　—Wilson Follet, *Modern American Usage*

Meaning is not what you start with, but what you
end up with.
　　—Peter Elbow

When a thought is too weak to be expressed
simply, it should be rejected.
　　—Vauven Argues

The objective of writing a first draft of your paper was to identify and organize
your ideas—to identify what you knew and, in the process, discover something

that you didn't know. Thus, you wrote the first draft primarily for yourself. Now it's time to rethink your ideas and communicate with your audience. To do this you must revise your work.

Learning how to revise your writing is the key to becoming a good writer. Revision is often hard work because it traps you in a double bind: You can't find the right words until you know exactly what you want to say, but you can't know exactly what you are saying until you find just the right words. Although revising what you've written is difficult, it is the essence of good writing. It involves rewriting *and rethinking* what you've already written (then rewriting what you've rewritten) while constantly and carefully studying every word, sentence, and paragraph to determine if they say *exactly* what you mean. Revising is essential because, while preparing a first draft, you often must write what you'll delete before discovering what you want to keep. Similarly, to know what you're trying to say you often must look at what you've written and ask if you've said it. Good writing is revision, not magic in producing a first draft. As experienced writers like to say, "There is no good writing, only good revision."[1]

Revising improves your writing and thinking by increasing the clarity and forcefulness of your message. You'll go over and over what you've written. All great writers do. For example, James Thurber's wife rated a first draft of one of Thurber's manuscripts as "high school stuff," to which Thurber replied, "Wait until the seventh draft."

People unwilling to revise and rethink their writing miss an important way to learn about their subject and improve communication with their audience. William Faulkner rewrote *The Sound and the Fury* five times, and Hemingway wrote the last page of *A Farewell to Arms* 39 times — until, as he later said, he got "the words right." Revising your writing is where the hard work begins. To again quote Hemingway, "Wearing down seven #2 pencils is a good day's work." You'll wear down most of your pencils when you start revising your work because, contrary to what some scientists believe, you cannot think of everything at once as you write.[2] What you write will constantly change and grow as your writing teaches you more about your subject. Therefore, think of your paper as a flexible framework rather than as a rigid box. This framework flexes most when you revise your paper.

---

[1]Inexperienced writers often balk at having to revise a paper. Nevertheless, such revisions are essential to learning about your subject and to producing an effective paper or essay. Moreover, the revision process mimics science. Scientists routinely repeat and revise their experiments to best attack hypotheses they're testing. For example, in 1941 George Beadle and E. L. Tatum used X-rays to form 1,000 mutated spore-cultures of the fungus *Neurospora crassa*, an orange bread-mold. They fed each of these cultures special diets, noting which diets promoted growth and which did not. In the 299th culture, vitamin B_6 (pyridoxine) restored proper growth. Beadle and Tatum concluded that radiation had damaged a gene responsible for making an enzyme needed for synthesis of vitamin B_6. From this work, Beadle and Tatum announced the single-gene single-enzyme hypothesis, for which they shared the 1958 Nobel Prize in physiology and medicine. Their work, which helped put modern genetics on a chemical basis, remains important for chemically controlling genetic diseases.

[2]An apparent exception to this was Rudolph Virchow, a German pathologist whose succinct statement, "every cell from a cell," completed the Cell Theory and repudiated theories of spontaneous generation. Virchow, we are told, did not revise his papers. His manuscripts, where still extant, show small corrections.

Rewriting Is Rethinking

*What makes me happy is rewriting. In the first draft
you get your ideas and your theme clear, if you
are using some kind of metaphor you get that
established, and certainly you have to know where
you're coming out. But the next time through it's like
cleaning house, getting rid of all the junk, getting
things in the right order, tightening things up. I like
the process of making writing neat.*
 —Ellen Goodman

Easy writing makes hard reading.
 —Ernest Hemingway

Many students view revising as a trivial, perfunctory clean-up involving little more than moving commas, while others view revising as punishment for not writing well the first time. Both of these attitudes inhibit your effectiveness because they ignore the power of writing as a learning tool. Revision improves your writing because *rewriting is rethinking*. No matter how mechanical, rewriting is critical to learning, understanding, and communicating.

You'll find many mistakes in the first draft of your paper. Some of these mistakes will be glaring: incomplete sentences, periods substituted for commas, and garbled paragraphs. Don't be alarmed by these mistakes. They're part of getting ideas out of your head and onto paper. Having done that with the first draft of your paper, what you must now do is correct the mistakes and produce a coherent document, both of which involve revising your work.

Revising your work is important for several reasons:

Revision improves learning. When done properly, revising your work reveals gaps and other problems that were disguised in your original paper by poor writing. Filling these gaps requires that you write new material. Moreover, revision often takes you to a point from which you can't proceed without either connecting ideas or eliminating material. Choosing between these options requires that you analyze the structure of your writing, which is synonymous with examining the structure of your scientific argument. This leads to reinvestigating your ideas, which extends beyond discovering and recording data to interpretation. When you revise, you rethink.

Revision helps you communicate better with your audience. When you wrote the first draft of your paper, you wrote to identify your ideas and discover new ideas. If you do not revise your work, all you'll have is a running account of your thinking. This may be interesting and valuable to you but probably will not impress others.

Revision helps you be objective. Your objectivity improves after you've set your paper aside for a few days to "let it cool."

Revision helps you scrutinize what you've written. You'll be able to polish, sharpen, correct, and focus your writing. More important, you'll be able to make your writing more accurate, precise, consistent, concise, and fair—all of the things that will help you learn more about your subject and help you communicate better with readers.

Although revising your writing is hard work, it is also rewarding. It is like furnishing a house that you built: You'll see the outline and general direction of the paper while molding your paper into a comfortable, readable form. Just as the remodeled house eventually becomes livable, so too will your paper communicate well with readers.

The principles described in the upcoming chapters will tell you how to revise your papers. You'll be unable to apply all of the principles at once, because removing one layer of errors often makes another layer appear. Thus, you should expect to revise your paper several times, until you've incorporated all of the principles discussed in the upcoming chapters. You'll apply many principles rather than make one or two major changes in the paper. Although any one of the changes that you make may seem insignificant, their cumulative effect determines whether you write well or write poorly. In writing, the victory goes to those who recognize and correct the many small problems that impede learning and communication.

To write well, you must learn to effectively revise your work. The first step is to delete clutter from your first draft. When you delete clutter, clear thinking becomes clear writing.

Deleting Clutter: Simplicity and Conciseness

If men would only say what they have to say in
plain terms, how much more eloquent they would be.
    —Samuel Coleridge, *On Style*

Writing improves in direct ratio to the number of
things we can keep out of it that shouldn't be there.
    —William Zinsser, *On Writing Well*

Brevity in writing is the best insurance for its perusal.
    —Rudolph Virchow

Everything should be made as simple as possible, but
not simpler.
    —Albert Einstein

Very simple ideas are within the reach of only very
complicated minds.
> —Remy De Gourmont

Beauty of style and harmony and grace and good
rhythm depend on simplicity.
> —Plato

Linus Pauling said, "Science is the search for truth." If you hide the truth with vague, pretentious clutter, you've not advanced science. To advance science and to do yourself justice, you must delete the clutter from your writing.

Many books about scientific writing urge scientists to "be brief." However, this rule is useless because effective writing is not synonymous with brief writing. Rather, effective writing is *concise* writing that takes the shortest time to be understood. Calvin Coolidge's wife learned the difference between brief and concise when she asked her husband what the preacher had talked about at church. "Silent Cal" replied, "Sin." When his wife then asked, "Well, what about it?" Calvin replied, "He was against it." Coolidge's responses were certainly brief, yet frustratingly incomplete and not concise.

To write clearly and concisely you must choose your words for minimum clutter and maximum strength. This first involves ruthlessly stripping your writing to its essentials by removing unnecessary words, pompous frills, and irrelevant details. All of this clutter is fuzz that annoys readers by obscuring and lessening the importance of your writing. Take Ernest Hemingway's crude but valuable advice: "The most essential gift for a good writer is a built-in shock-proof shit-detector."

Clutter appears in all kinds of writing and is especially common when people try to overstate simple ideas. Educators are notorious for this. Consider this example:

> Realization has grown that the curriculum or the experiences of learners change and improve only as those who are most directly involved examine their goals, improve their understandings and increase their skills in performing the tasks necessary to reach newly defined goals. This places the focus upon the teacher, lay citizen and learner as partners in curricular improvement and as the individuals who must change, if there is to be curriculum change.

When you run this through your clutter decoder, you get something like this:

> Teachers, parents, and students must help if we are to change the curriculum.

Notice that the fog caused by the original version was not due to technical words; it was due merely to overstating a simple idea.

Tightening your writing into fewer and shorter words saves space and readers' time while making your message more obvious and powerful. However, the trick is not merely to delete words—anyone can do that—but to know what words to delete and why. Delete words that contribute nothing to what you're trying to say. Just as pruning and removing dead limbs can help a tree grow taller, so too can deleting unnecessary words strengthen a paper. However, just as too much pruning kills a tree, so too can too much deleting weaken your writing. Therefore, strive not for mere brevity, but rather for *conciseness*— brevity *and* completeness. To do this, get rid of the fuzz in your writing. A compulsion to keep everything means not that you write well, but merely that you lack discrimination.

Although writing can be too compact and terse, wordiness is a more common problem. This is because it is harder to write a concise paper than a wordy one. As American humorist Ambrose Bierce once scribbled at the end of a letter to a friend, "Forgive me for writing such a long letter. I didn't have time to write a short one."

Whenever Possible, Use Simple Words and Sentences

A man of true science . . . uses but few hard
words, and those only when none other will answer
his purpose; whereas the smatterer in science
. . . thinks that by mouthing hard words, he proves
that he understands hard things.
    —Herman Melville

Never fear big words. Long words name little things.
All big things have little names, such as Life and
death, Peace and war, or dawn, day, night, live,
home. Learn to use little words in a big way—it is
hard to do. But they say what you mean. When you
don't know what you mean, use big words: They
often fool little people.
    —*SSC BOOKNEWS*, July 1981

Writing is easy. All you have to do is cross out the
wrong words.
    —Mark Twain

A poorly built sentence of twenty words is about
four times as hard to attend to and to understand as
an equally poor sentence of ten words.
    —Walter Pitkin

Small words are the best, and the old small words
are best of all.
    —Winston Churchill

Unavoidable Big Words

Bertrand Russell, when asked about the relationship of simple words to effective writing, said that, "Big men write little words, little men write big words." Although this is usually true in popular writing, it's not always true when writing about biology. For example, there are no short, simple words to replace big words such as *endoplasmic reticulum, photosynthesis,* or *chromatography.* Other biological words are even bigger. For example, the scientific name of the Dahlemense strain of tobacco mosaic virus has 1,185 letters, that of bovine glutamate dehydrogenase has 3,641 letters, and the name of tryptophane synthetase A (a listing of 267 amino acids) has 1,913 letters. Similarly, the longest word in the *Oxford English Dictionary Supplement* is *pneumonoultramicroscopic-silicovolcanoconiosis,* a 45-letter monstrosity that refers to a lung disease gotten by some miners. Coming in at a close second is *hepaticocholangiocholecysten-terostomies,* a medical term referring to the surgical creation of new communications between gallbladders and hepatic ducts and between intestines and gallbladders. However, even these words seem like runts when compared to the systematic name for DNA of the human mitochondria that contains 16,569 nucleotide sequences (published in key form in the April 9, 1981 issue of *Nature*)—a word that has about 207,000 letters.

Alice had not the slightest idea what Latitude was,
or Longitude either, but she thought they were nice
grand words to say.
    —Lewis Carroll, *Alice's Adventures in*
        Wonderland

Many biologists are like Alice: They insist on using as many nice, grand words as possible, even when those words are unnecessary and when the writer is unsure of the words' meanings. Consequently, these writers often write tedious, incomprehensible papers and leave themselves open to the reprimand that Alice received: "'Speak English!' said the Eaglet. 'I don't know the meaning of half those long words, and, what's more, I don't believe you do either!'"

Simple writing communicates well and indicates clear thinking. Consequently, the most effective writing uses short, simple words and phrases. For example, Lincoln's 701-word second inaugural address contained 505 one-syllable words and 122 two-syllable words. Similarly, some of the greatest scientists of all time, including Louis Pasteur, Albert Einstein, and Marie Curie, explained their complex ideas with simple words that other scientists and the public could understand and appreciate. Such scientists knew that simple writing communicates more effectively than complex writing. They also knew that their clear, simple writing contributed as much to their stature and fame as did their brilliant ideas and discoveries.

(*Text continued on page 74.*)

Proofreader's Marks

You'll speed your editing by using proofreader's marks, a set of symbols that save you from having to repeat your editing instructions.

Problem	Symbol	Example
Omitted word	∧ (caret)	study describes ∧ effects *the*
Omitted letter	∧ (caret)	that b∘ok
Transpose letters	∼	f∂rm the ocean
Transpose words	⌒	was only exposed
Capitalize word	☰ (three short underlines)	these data ☰
Word should be lowercase	╱ (slash)	⟋hese data
Italicize	▬ (underline once)	Homo sapiens
Separate words	⎮ (draw vertical line between)	read⎮carefully
Delete word	▬▬ (draw line through)	the ~~nice~~ data
Delete letter	╱ (draw line through)	the nic⟋e data
Do not leave a space	⌒ (sideways parentheses)	the e⁀nd
Wrong letter	╱ (draw line through and add correct letter above)	*f* ⟋emale
Wrong word	▬▬ (draw line through and add correct word above)	*These* ~~This~~ data
Start a new paragraph	¶ (paragraph symbol)	female. ¶ In contrast
Let it stand	(stet)	The energy ~~needs~~ *stet*
Insert comma	⌃ (caret)	red⌄small⌄and poisonous

Let's apply these marks to this essay about anabolic steroids.

anabolic sterods function marvelously to develope maintaain ma le sed characteristics. We definitely know that they are synthetic derivatives of Testosterone because we have made them. They not only effect one phyiscally, but also psychologically and mentally. Some reviews of the subject suggest that anabolic steroids may even have an addictive potential similar to the many, many other drugs that are abused everyday by people. Anabolic steroids are psychoactive compounds, as evidenced by their well-documented affects on behavior. Neuronal anabolic steroid receptors have been identified in the brain, suggesting the neurochemical basis for their psychoactive effects. Most addicts suffer symptoms withdrawal and headaches was unsuccessful in his efforts to cut down on use, and continued to use the steroids despite their knowledge that he was having emotional and martial problems related to their use. These were just a few of examples to emphasize the possibility of psychoactive substance dependence. In conclusion, their patients developed a dependence on anabolicsteroids that was strikingly similar to dependencies seen with other drugs.

Here's how this essay looks after revision:

anabolic sterods ~~function marvelously to~~ are synthetic derivatives of testosterone, a hormone that develope maintains male sed in

characteristics. ~~We definitely know that they are synthetic derivatives of~~

Anabolic steroids affect people ~~Testosterone because we have made them. They not only effect one~~ physically,

and may be as ~~but also psychologically~~ and mentally, ~~Some reviews of the subject suggest~~

as drugs such as ~~that anabolic steroids may even have an~~ addictive ~~potential similar to the~~

cocaine. Indeed, our brains have receptors for ~~many, many other drugs that are abused everyday by people.~~ Anabolic steroids

~~are psychoactive compounds, as evidenced by their well-documented affects~~

¶ Addiction to anabolic steroids resembles that to other ~~on behavior. Neuronal anabolic steroid receptors have been identified in the~~

drugs. For example, when deprived of anabolic steroids, ~~brain, suggesting the neurochemical basis for their psychoactive effects.~~ Most

such as   and were   their addicts suffer symptoms withdrawal and headaches. ~~was unsuccessful in his~~

However, most addicts do not reduce their use of the ~~efforts to cut down on use, and continued to use the steroids despite their~~

drug, despite knowing the harm it causes. ~~knowledge that he was having emotional and martial problems related to their~~

~~use. These were just a few of examples to emphasize the possibility of~~

~~psychoactive substance dependence. In conclusion, their patients developed a~~

~~dependence on anabolic steroids that was strikingly similar to dependencies~~

~~seen with other drugs.~~

(Continued on following page)

Here's the product:

Anabolic steroids are synthetic derivatives of testosterone, a hormone that develops and maintains sex characteristics in males. Anabolic steroids affect people physically and mentally, and may be as addictive as drugs such as cocaine. Indeed, our brains have receptors for anabolic steroids.

Addiction to anabolic steroids resembles that to other drugs. For example, when deprived of anabolic steroids, most addicts suffer withdrawal symptoms such as headaches. However, most addicts do not reduce their use of the drug, despite knowing the harm it causes.

If you want to communicate with your readers, use simple words and phrases. This produces concise writing, which can carry a tremendous punch. For example, the Lord's Prayer has only 56 words, the Twenty-third Psalm 118 words, the Gettysburg Address 258 words (196 of which are one-syllable), Newton's First Law of Motion 29 words, the First Amendment to the U.S. Constitution 45 words, the Ten Commandments only 75 words, and the Golden Rule only 11 words. For comparison, the U.S. Department of Agriculture's directive for pricing cabbage contains 15,629 words. You can guess which of these documents is hardest to read.

Just as simple words such as *life, death, love,* and *earth* describe profound concepts, simple sentences and paragraphs often convey much wisdom. Prefer the simple, understandable word or sentence to one that is stilted and long-winded. For example, writing that "The biota in question in this particular study exhibited a 100% mortality rate" wastes readers' time and shows that the writer is either unwilling or incapable of thinking clearly. Instead, write, "All of the rats died." Aim at functional beauty, not superficial grace, and constantly ask what can be deleted, shortened, or simplified in your writing. Simplicity is important in science, especially when stating conclusions for the first time. Also, as you'll learn in Chapter 5, simplicity has an added benefit: It simplifies punctuation.

I can't list all of the ways that scientists clutter their writing. However, the most common causes and solutions are discussed here.

Delete Unnecessary Words

If it is possible to cut a word out, always cut it out.
    —George Orwell

Each unnecessary word or phrase, no matter how small, tires readers and lessens their receptivity to your ideas. Therefore, delete windy phrases and "bankrupt" words that, rather than adding information, only delay readers from grasping your ideas. For example, consider deleting these words and phrases:

As you know	(If the readers know what you're about to say, don't waste their time by re-telling them the information.)
Needless to say	(If it's needless to say, then don't say it.)
I might add that	(Just add it.)
It is important to state here that	(This suggests that other material that you've included may be unimportant.)
It is worth noting that	(This suggests that other things you've said are not worth noting)
When time permits	(This may sound poetic, but it's inaccurate. Time cannot permit anything.)
As already stated	(If you've already said it, don't say it again.)
It goes without saying that	(If it goes without saying, let it go without saying.)

From this point of view, it is relevant to mention that

It is of interest to note that

In order to keep the problem in perspective, we would like to emphasize that

It is considered, in this connection, that

It appears that

As a matter of fact

as such

for the sake of

in the case of

in some instances

take this opportunity

the fact that

the reason is

a distance of

a period of

As of this date

in a very real sense

in my opinion

surely

doubtless

quite

As a matter of fact,

It stands to reason that

It has been shown that

It has long been known that

It has been demonstrated that

It must be remembered that

It may seem that

It is worthy to note that

It is clear that

It may be mentioned that

It is worth pointing out that

It is interesting to note that

It is significant to note that

It is relevant to mention here that

It is well known that

It should be kept in mind in this
particular connection that

It is clearly obvious from data in
Fig. 2 that

It has come to our attention that

It may be said that

For all intents and purposes

generally

virtually

actually

really

kind of

definitely

a lot

perhaps

really

sort of

very

practically

certain

given

Redundant words and phrases are also unnecessary. Lawyers are notorious for the poor writing that accompanies their use of redundant phrases such as *null and void, aid and abet, sum and substance,* and *irrelevant and immaterial, cease and desist,* and *give and convey*—all of which are often buried in a blizzard of *whereases* and *hereinafters.*[3] Although redundancy is essential for safety, it is awful in scientific writing. In writing, redundancies usually appear as tautologies, which are phrases that say the same thing twice using different

---

[3]For an excellent discussion of how lawyers' poor writing creates circular definitions, see Gopen, George D. 1987. "Let the Buyer in the Ordinary Course of Business Beware: Suggestions for Revising the Prose of the Uniform Commercial Code." *The University of Chicago Law Review* 54: 1178–1214.

words. Like stuttering in speech, redundancy in writing detracts from communication. Don't confuse restatement with explanation, and don't cover the same ground twice. Say something once, and say it well.

Delete redundant words and phrases such as:

~~viable~~ alternative	two ~~different~~ methods
~~ultimate~~ outcome	~~absolutely~~ essential
~~past~~ medical history	9 a.m. ~~in the morning~~
~~young~~ juvenile	~~advance~~ notice
~~authentic~~ replica	~~general~~ consensus
period ~~of time~~	~~absolutely~~ complete
ask ~~the question~~	assembled ~~together~~
~~total of~~ 41 people	~~exactly~~ the same
~~previously~~ found	~~in close~~ proximity
~~advance~~ planning	after ~~the conclusion of~~
by ~~means of~~	due to ~~the fact that~~
hurry ~~up~~	~~in order~~ to
~~end~~ product	near ~~the place of~~
combine ~~together~~	~~completely~~ unanimous
consensus ~~of opinion~~	cooperate ~~together~~
each ~~and every~~	endorse ~~on the back~~
~~entirely~~ eliminated	~~final~~ outcome
few ~~in number~~	~~personal~~ friend
~~personal~~ opinion	recur ~~again~~
reduce ~~down~~	resume ~~again~~
separate ~~out~~	~~in connection~~ with
disappear ~~from sight~~	spherical ~~in shape~~
~~come to an~~ end	~~science of~~ biology
for ~~the purpose of~~	plan ~~ahead for the future~~
~~duly~~ noted	~~wish to~~ thank
~~mutual~~ cooperation	~~most~~ unique
red ~~in color~~	introduce ~~a new~~

~~already~~ existing

~~currently~~ underway

never ~~before~~

~~continue to~~ remain

~~first~~ began

mix ~~together~~

~~private~~ industry

~~the question as to~~ whether

basic ~~and fundamental~~

~~various~~ differences

~~each~~ individual

if ~~at all~~ possible

~~perform a~~ study

join ~~together~~

slow ~~up~~

hectares ~~of land~~

~~all~~ throughout

large ~~in size~~

~~exactly~~ identical

~~underlying~~ purpose

~~viable~~ solution

~~past~~ experience

~~joint~~ partnership

while ~~at the same time~~

~~at the time~~ when

~~and so on~~ and so forth

~~completely~~ finish

each ~~individual~~

~~initial~~ preparation

~~absolute~~ necessity

~~basic~~ fundamentals

~~completely~~ eliminate

~~separate~~ entities

~~currently~~ being

had done ~~previously~~

none ~~at all~~

~~joint~~ cooperation

any ~~and all~~

~~completely~~ finish

~~future~~ plans

~~unexpected~~ surprise

bisect ~~into two parts~~

~~one and the~~ same

blame ~~it on~~

~~actual~~ facts

~~basic~~ essentials

~~definite~~ decision

~~usual~~ custom

subject ~~matter~~

~~equally~~ as effective

~~two equal~~ halves

~~have~~ need ~~for~~

~~any and~~ all

~~full and~~ complete

~~various~~ different

near ~~the vicinity of~~

unusual ~~in nature~~

~~active~~ consideration

~~baffling~~ enigma

~~conclusive~~ proof

~~advance~~ reservation

~~close~~ proximity

~~basic~~ fundamentals

brief ~~in duration~~

merge ~~together~~

repeat ~~the same~~

until ~~such time as~~

the ~~actual~~ number

~~conclusive~~ proof

stunted ~~in growth~~

~~hard~~ evidence

assemble ~~together~~

during ~~the course of~~

revert ~~back~~

~~advance~~ plan

~~close~~ proximity

~~current~~ status

~~honest~~ truth

~~overall~~ plan

repeat ~~again~~

this ~~particular~~ instance

whether ~~or not~~

by ~~means of~~

balance ~~against one another~~

range ~~all the way~~ from

add ~~an additional~~

during ~~the course of~~

~~excess~~ verbiage

~~mutual~~ cooperation

refer ~~back~~

because ~~of the fact that~~

nominated for ~~the position of~~

~~make a~~ study ~~of~~

~~absolutely~~ essential

~~completely~~ surround

~~deliberately~~ chosen

~~quite~~ impossible

~~wholly~~ new

~~customary~~ practice

~~past~~ experience

rarely ~~ever~~

any ~~and all~~

combine ~~into one~~

~~final~~ outcome

one ~~and the same~~

first ~~and foremost~~

refer ~~back~~ to

~~different~~ species

all ~~of~~

~~uniformly~~ consistent

cancel ~~out~~

for ~~the purpose of~~

continue ~~on~~

~~early~~ beginnings

~~necessary~~ requisite

debate ~~about~~

repeat ~~again~~

~~in~~ between

~~still~~ remain

circulate ~~around~~

file ~~away~~	join ~~together~~
protrude ~~out~~	blend ~~together~~
write ~~up~~	is ~~defined as~~
~~close~~ scrutiny	~~completely~~ full
~~consequent~~ results	~~equal~~ halves
~~definite~~ proof	~~end~~ result
enclosed ~~herewith~~	if ~~it is assumed that~~
~~entirely~~ eliminate	~~at a~~ later ~~date~~
~~in conjunction~~ with	smaller ~~in size~~
by ~~means of~~	~~in~~ between
~~serious~~ crisis	~~new~~ initiatives
subject ~~matter~~	mix ~~together~~

Avoid placing adjectives before absolute words such as *dead, extinct, unique,* and *final* that resist modifiers. Also avoid throat-clearing words such as *basically* and *ideally.* These words add nothing to a sentence's meaning. Similarly, avoid tagging nouns with words that identify those nouns. For example:

Our study was done 100 km south of ~~the city of~~ Chicago, Illinois, during ~~the month of~~ April. The scientists included ~~a group of~~ biologists.

Also watch for redundant nouns such as *size* (large ~~in size~~), *color* (green ~~in color~~), *weight* (heavy ~~in weight~~), *process* (~~the process of~~ cellular respiration), *concept* (~~the concept of~~ biology), and *number* (few ~~in number~~). Write that a mucous membrane is pink, not that it is pink *in color.* Be a "word-miser," but don't hesitate to use a longer word, phrase, or sentence when it makes your writing more clear, more precise, or more accurate. Also watch for words such as *phenomenon, virtually, element, objective, primary,* and *constitute*; some scientists use these words to dress up a simple statement or to give an air of impartiality to a biased judgment.

Rid your writing of pretentious, pseudointellectual words such as *utilize* and *facilitate*: Just say *use* and *help.* If you insist on using pretentious, bloated language, your writing will resemble a failing dieter who can't resist second helpings and desserts, neither of which help to achieve the goal of losing weight. Biologists who can't resist inserting unnecessary phrases, such as, "*It is interesting to note that,*" produce fat, lifeless papers that few scientists will want to read and even fewer editors will want to publish.

Many scientists insist on using Latin or Greek derivations of Anglo-Saxon words rather than the simpler Anglo-Saxon words themselves. These folks write sentences such as, "I prefer an abbreviated phraseology, distinguished for its lucidity," when all they mean is, "I like short, clear words." Prefer the short, simple word. *Avoid hippopotomonstrosesquipedalian construction of written prose.* To write effectively, avoid ~~excess~~ verbiage. For example,

POMPOUS: Prices were impacted adversely by the season's aridity, which was deleterious to agriculture.

IMPROVED: The dry summer hurt the crops and increased prices.

Replace Large Words and Phrases with Simple Words

A word may be a fine-sounding word, of an unusual
length, and very imposing from its learning and
novelty, and yet in the connection in which it is
introduced, may be quite pointless and irrelevant.
    —William Hazlitt

In general those who have nothing to say contrive
to spend the longest time doing it.
    —James Russell Lowell

The pseudo prestige of long and difficult words
transcends the useful scientific term and diffuses
widely through our papers. Simple things are made
complicated, and the complex is made incomprehensible. Chaos reigns. The so-called medical
literature is stuffed to bursting with junk, written in a
hopscotch style characterized by a Brownian
movement of uncontrolled parts of speech which
seethe in restless unintelligibility.
    —William B. Bean

Most of the fundamental ideas of science are
essentially simple, and may, as a rule, be expressed
in a language comprehensible to everyone.
    —Albert Einstein

Many scientists use wordy and excessively formal language either to show others how smart they are or to compensate for their professional shortcomings. These writers write to impress rather than to express and therefore ignore their responsibility of communicating with readers.

You'll communicate most effectively, and therefore impress your readers most favorably, if you choose simple, direct words over long-winded, preten-tious words. Furthermore, expressing your ideas in simple terms is a sure way to determine if your reasoning is logical. If it is, your logic will be obvious when you simplify your writing. If it isn't, that too will be clear, and you may want to leave the clutter in your writing as a way of covering your lack of thought and logic. Instead of relying on cluttered, poor writing to deceive readers, rethink and rewrite what you've written. Poor writing only diverts attention from your ideas.[4]

Here are some long-winded words and phrases that can be replaced by simpler words (in the left column):

about	on the order of; as regards; with regard to; concerning the matter of; in regard to; approximately; in the neighborhood of; in the approximate amount of; in reference to
adjust	make an adjustment
after	subsequent to; at the conclusion of; following
agree	are found to be in agreement; are of the same opinion
allow	afford an opportunity to
also	additionally
although	despite the fact that; notwithstanding the fact that
always	in all cases
analyze	perform an analysis of
apparently	it is apparent that
are	have been shown to be
as	in view of the fact that
ask	inquire; request
aware	cognizant of
because	due to the fact that; as a consequence of; as a result of; in light of the fact that; owing to the fact that; because of the fact that; on account of the fact that; inasmuch as; on the grounds that; prior to this point in time; accounted for by the fact
before	in advance of; prior to
begin	initiate; commence
build	construct

---

[4]Many authors divert readers' attention by using unnecessarily complex writing. Siegfried (Siegfried, John J. 1970. A first lesson in econometrics. *Journal of Political Economy* 78: 1378–1379) humorously attacked this strategy with a "rigorous" mathematical demonstration that $1 + 1 = 2$.

can	has the ability to; is able to; is in a position to; has the opportunity to; has the capacity for
cannot	not in a position to
cause	give rise to
change	modification
claim	allege
consider	take into consideration
copy	duplicate
cut	incision
decide	arrive at a decision
describe	give an account of
despite	in spite of the fact that
do	perform; accomplish; achieve; implement; carry out
during	in the course of
early	ahead of schedule
ease	facilitate
education	educational process
end	terminate; conclude; finalize
enough	sufficient
examine	make an examination of
exceeds	in excess of
except	with the exception of; with the only difference being
expect	anticipate
explain	elucidate
extra	superfluous; additional
face	confront
few	small number of
find	ascertain the location of; locate
finish	bring to a conclusion
first	initial; the thing to do before anything else

for	in the amount of; on behalf of
for example	an example of this is the fact that
free	disengage
frequently	is often the case that
full	replete
fully	to the fullest possible extent
get	procure; obtain
give	donate; contribute
go	proceed
happen	eventuate
have	am in possession of
help	facilitate; assist; render assistance to
idea	concept
if	in the event that; assuming that; under conditions in which; in case
important	of great importance
in	during the month of
inaccurate	not a high order of accuracy
inhibited	produced an inhibitory effect
instead	in lieu of
investigate	conduct an investigation of
is	has been shown to be; is widely acknowledged to be
judge	adjudicate
know	be cognizant of
later	subsequently
like	along the lines of; similar in character to
list	enumerate
long	lengthy
lost weight	experienced a weight loss
many	a considerable number of; a large number of

may	the chance that; is possible that; could happen that
measure	quantify
meet	come in contact with; make the acquaintance of; encounter
methods	methodology
most	the majority of; the predominant number of
much	a considerable amount of; a great deal of; quite a large quantity of
must	is necessary that
named	by the name of
near	in the vicinity of; in close proximity to
next	subsequent
never	in no case; on no occasion
notice	take cognizance of
now	at the present time; at this point in time; presently; in this day and age
often	in many cases
opposite	antithesis
part	component
perhaps	it may well be that
please	I would appreciate it if you would; I would like to ask that
propose	make a proposal
previously	at an earlier date
purify	achieve purification
rapidly	at rapid rate
rarely	in rare cases; in only a very small number of cases
regularly	on a regular basis
remind	call attention to the fact that
replaces	supersedes
require	involve the necessity of

result	resultant effect
risk	jeopardize
satisfactorily	in a satisfactory manner
say	assert
saw	was witness to
several	a number of
show	demonstrate
some	a number of
sometimes	in some cases
soon	in the not-too-distant future
start	initiate
stress	place a major emphasis on
study	undertake an examination of
surgery	surgical procedure
then	after this has been done
tiny	minuscule
try	endeavor
twice	on two separate occasions
use	utilize
were	proved to be
while	during the time that
without	in the absence of
yearly	on an annual basis

Do not use imaginary words such as *analyzation, interpretate, irregardless,* and *unequivocably.* Replace these nonexistent words with *analysis, interpret, regardless,* and *unequivocally.* Also remember that *inflammable* and *flammable* both mean *combustible.* If you want to say that something will not burn, just say that it's *nonflammable* or *noncombustible.*

Don't let your writing sound like Dr. Ray Stantz (played by Dan Akroyd) of *Ghostbusters.* Ray is bright but always babbles about the latest scientific gizmos: "If the ionization rate is constant for all ectoplasmic entities, we could really bust some heads—in a spiritual sense." Conversely, Dr. Peter Venkman (Bill

Murray) is someone like the rest of us and has no idea what phrases such as "total protonic reversal" or "PKE valences" mean. Peter tells Ray, ". . . just tell me what the hell is going on."

Take Peter's advice. Just tell readers what is going on. You'll do that best by simplifying your writing.

Simplify Your Writing

If you would be pungent, be brief; for it is with words as with sunbeams — the more they are condensed the deeper they burn.
—Robert Southbey

It is the essence of genius to make use of the simplest ideas.
—Charles Péguy

Genius is the ability to reduce the complicated to the simple.
—C.W. Ceram

On the whole, I think the pains which my father took over the literary part of the work was very remarkable. He often laughed or grumbled at himself for the difficulty which he found in writing English, saying, for instance, that if a bad arrangement of a sentence was possible, he would be sure to adopt it. . . . When a sentence got hopelessly involved, he would ask himself "now what do you want to say?" and his answer written down, would often disentangle the confusion.
—Francis Darwin, The Life of Charles Darwin

Just as unneeded pounds slow the movement of an overweight person, so too do excess words rob your writing of vitality by obscuring your ideas. Verbiage inhibits communication because the more a reader has to concentrate on, the less attention he or she can give to an idea. However, simple writing can also be flat and dry. The product resembles unsalted meat and potatoes: edible, but hardly memorable. To produce memorable writing that communicates effectively, you must make your writing clear and concise.

Write as simply and concisely as you can. Although wordiness is not synonymous with redundancy, it is a first cousin. Moreover, it usually accompanies shallow thought or poor understanding. Therefore, don't be a windbag: Don't use 20 words to say what you can with 15. Remember that there's no direct relationship between the length of what you write and its significance. For example, Watson and Crick's paper in the April 1953 issue of *Nature* describing the structure of DNA was only one page long, yet it revolutionized biology.

Simplify your writing by asking yourself what can be eliminated from your writing. Use the least number of words to say exactly what you mean in a way that cannot be misunderstood.

Rachel Carson

The Obligation to Endure

Few books have had a greater impact than Rachel Carson's (1907–1964) *Silent Spring*, one of the most important books of the twentieth century. Carson wrote *Silent Spring* in response to a letter that she received in January 1958 from Olga Huckins describing how a small part of the world had been made lifeless by pesticides. Carson's response was that "There would be no peace for me if I kept silent."

Carson, who had spent most of her professional life as a writer and marine biologist with the U.S. Bureau of Fisheries (now the U.S. Fish and Wildlife Service), began to study the use of pesticides such as DDT. Four years later, she published *Silent Spring*. Most magazines, fearing lost advertising income, refused to publish excerpts of the book, despite Carson's fame for writing *The Sea Around Us* several years earlier.[5] Chemical companies spent hundreds of thousands of dollars to block publication, and one manufacturer of canned baby food even claimed that her book would cause "unwarranted fear" to mothers who used their product. The book was attacked as scientifically unsound and emotionally motivated, and Carson was maligned as "ignorant" and "hysterical." Nevertheless, Carson's message and brilliant writing made *Silent Spring* a best-seller. Moreover, it had a tremendous impact on the world. For example, *Silent Spring* prompted then President Kennedy to create a special panel of the Science Advisory Committee to study the use of pesticides. That panel's report vindicated Carson and stimulated the government to act against pollution. *Silent Spring* helped launch the movement that made "ecology" a household word.

---

[5]The only exception was *The New Yorker*, which serialized *Silent Spring* before the book's publication.

Here is "The Obligation to Endure" from *Silent Spring*. This chapter uses imaginative fiction and scientific exposition to persuade readers.

The history of life on earth has been a history of interaction between living things and their surroundings. To a large extent, the physical form and the habits of the earth's vegetation and its animal life have been molded by the environment. Considering the whole span of earthly time, the opposite effect, in which life actually modifies its surroundings, has been relatively slight. Only within the moment of time represented by the present century has one species—man—acquired significant power to alter the nature of his world.

During the past quarter century this power has not only increased to one of disturbing magnitude but it has changed in character. The most alarming of all man's assaults upon the environment is the contamination of air, earth, rivers, and sea with dangerous and even lethal materials. This pollution is for the most part irrecoverable; the chain of evil it initiates not only in the world that must support life but in living tissues is for the most part irreversible. In this now universal contamination of the environment, chemicals are the sinister and little-recognized partners of radiation in changing the very nature of the world—the very nature of its life. Strontium 90, released through nuclear explosions into the air, comes to earth in rain or drifts down as fallout, lodges in soil, enters into the grass or corn or wheat grown there, and in time takes up its abode in the bones of a human being, there to remain until his death. Similarly, chemicals sprayed on croplands or forests or gardens lie long in soil, entering into living organisms, passing from one to another in a chain of poisoning and death. Or they pass mysteriously by underground streams until they emerge and, through the alchemy of air and sunlight, combine into new forms that kill vegetation, sicken cattle, and work unknown harm on those who drink from once-pure wells. As Albert Schweitzer has said, "Man can hardly even recognize the devils of his own creation."

It took hundreds of millions of years to produce the life that now inhabits the earth—eons of time in which that developing and evolving and diversifying life reached a state of adjustment and balance with its surroundings. The environment, rigorously shaping and directing the life it supported, contained elements that were hostile as well as supporting. Certain rocks gave out dangerous radiation; even within the light of the sun, from which all life draws its energy, there were short-wave radiations with power to injure. Given time—time not in years but in millennia—life adjusts, and a balance has been reached. For time is the essential ingredient; but in the modern world there is no time.

The rapidity of change and the speed with which new situations are created follow the impetuous and heedless pace of man rather than the deliberate pace of nature. Radiation is no longer merely the background radiation of rocks, the bombardment of cosmic rays, the ultraviolet of the sun that have existed before there was any life on earth; radiation is now the unnatural creation of man's tampering with the atom. The chemicals to which life is asked to make its adjustment are no longer merely the calcium and silica and copper and all the rest of the minerals washed out of the rocks and carried in rivers to the sea; they are the synthetic creations of man's inventive mind, brewed in his laboratories, and having no counterparts in nature.

To adjust to these chemicals would require time on the scale that is nature's; it would require not merely the years of a man's life but the life of generations. And even this, were it by some miracle possible, would be futile, for the new chemicals come from our laboratories in an endless stream; almost five hundred annually find their way into actual use in the United States alone. The figure is staggering and its implications are not easily grasped—500 new chemicals to which the bodies of men and animals are required somehow to adapt each year, chemicals totally outside the limits of biologic experience.

Among them are many that are used in man's war against nature. Since the mid-1940's over 200 basic chemicals have been created for use in killing insects, weeds, rodents, and other organisms described in the modern vernacular as "pests"; and they are sold under several thousand different brand names.

These sprays, dusts, and aerosols are now applied almost universally to farms, gardens, forests, and homes—nonselective chemicals that have the power to kill every insect, the "good" and the "bad," to still the song of birds and the leaping of fish in the streams, to coat the leaves with a deadly film, and to linger on in soil—all this though the intended target may be only a few weeds or insects. Can anyone believe it is possible to lay down such a barrage of poisons on the surface of the earth without making it unfit for all life? They should not be called "insecticides," but "biocides."

The whole process of spraying seems caught up in an endless spiral. Since DDT was released for civilian use, a process of escalation has been going on in which ever more toxic materials must be found. This has happened because insects, in a triumphant vindication of Darwin's principle of the survival of the fittest, have evolved super races immune to the particular insecticide used, hence a deadlier one has always to be developed—and then a deadlier one than that. It has happened also because, for reasons to be described later, destructive insects often undergo a "flareback," or resurgence, after spraying, in numbers greater than before. Thus the chemical war is never won, and all life is caught in its violent crossfire.

Along with the possibility of the extinction of mankind by nuclear war, the central problem of our age has therefore become the contamination of man's total environment with such substances of incredible potential for harm—substances that accumulate in the tissues of plants and animals and even penetrate the germ cells to shatter or alter the very material of heredity upon which the shape of the future depends.

Some would-be architects of our future look toward a time when it will be possible to alter the human germ plasm by design. But we may easily be doing so now by inadvertence, for many chemicals, like radiation, bring about gene mutations. It is ironic to think that man might determine his own future by something so seemingly trivial as the choice of an insect spray.

All this has been risked—for what? Future historians may well be amazed by our distorted sense of proportion. How could intelligent beings seek to control a few unwanted species by a method that contaminated the entire environment and brought the threat of disease and death even to their own kind? Yet this is precisely what we have done. We have done it, moreover, for reasons that collapse the moment we examine them. We are told that the enormous and expanding use of pesticides is necessary to maintain farm production. Yet is our real problem not one of *overproduction*? Our farms, despite measures to remove acreages from production and to pay farmers *not* to produce, have yielded such a staggering excess of crops that the American taxpayer in 1962 is paying out more than one billion dollars a year as the total carrying cost of the surplus-food storage program. And is the situation helped when one branch of the Agriculture Department tries to reduce production while another states, as it did in 1958, "It is believed generally that reduction of crop acreages under provisions of the Soil Bank will stimulate interest in use of chemicals to obtain maximum production on the land retained in crops."

All this is not to say there is no insect problem and no need of control. I am saying, rather, that control must be geared to realities, not to mythical situations, and that the methods employed must be such that they do not destroy us along with the insects.

The problem whose attempted solution has brought such a train of disaster in its wake is an accompaniment of our modern way of life. Long before the age of man, insects inhabited the earth—a group of extraordinarily varied and adaptable beings. Over the course of time since man's advent, a small percentage of the more than half a million species of insects have come into conflict with human welfare in two principal ways: as competitors for the food supply and as carriers of human disease.

Disease-carrying insects become important where human beings are crowded together, especially under conditions where sanitation is poor, as in time of natural disaster or war or in situations of extreme poverty and deprivation. Then control of some sort becomes necessary. It is a sobering fact, however, as we shall presently see, that the method of massive chemical control has had only limited success, and also threatens to worsen the very conditions it is intended to curb.

Under primitive agricultural conditions the farmer had few insect problems. These arose with the intensification of agriculture—the devotion of immense acreages to a single crop. Such a system set the stage for explosive increases in specific insect populations. Single-crop farming does not take advantage of the principles by which nature works; it is agriculture as an engineer might conceive it to be. Nature has introduced great variety into the landscape, but man has displayed a passion for simplifying it. Thus he undoes the built-in checks and balances by which nature holds the species within bounds. One important natural check is a limit on the amount of suitable habitat for each species. Obviously then, an insect that lives on wheat can build up its population to much higher levels on a farm devoted to wheat than on one in which wheat is intermingled with other crops to which the insect is not adapted.

The same thing happens in other situations. A generation or more ago, the towns of large areas of the United States lined their streets with the noble elm tree. Now the beauty they hopefully created is threatened with complete destruction as disease sweeps through the elms, carried by a beetle that would have only limited chance to build up large populations and to spread from tree to tree if the elms were only occasional trees in a richly diversified planting.

Another factor in the modern insect problem is one that must be viewed against a background of geologic and human history: the spreading of thousands of different kinds of organisms from their native homes to invade new territories. This worldwide migration has been studied and graphically described by the British ecologist Charles Elton in his recent book *The Ecology of Invasions*. During the Cretaceous Period, some hundred million years ago, flooding seas cut many land bridges between continents and living things found themselves confined in what Elton calls "colossal separate nature reserves." There, isolated from others of their kind, they developed many new species. When some of the land masses were joined again, about 15 million years ago, these species began to move out into new territories—a movement that is not only still in progress but is now receiving considerable assistance from man.

The importation of plants is the primary agent in the modern spread of species, for animals have almost invariably gone along with the plants, quarantine being a comparatively recent and not completely effective innovation. The United States Office of Plant Introduction alone has introduced almost 200,000 species and varieties of plants from all over the world. Nearly half of the 180 or so major insect enemies of plants in the United States are accidental imports from abroad, and most of them have come as hitchhikers on plants.

In new territory, out of reach of the restraining hand of the natural enemies that kept down its numbers in its native land, an invading plant or animal is able to become enormously abundant. Thus it is no accident that our most troublesome insects are introduced species.

These invasions, both the naturally occurring and those dependent on human assistance, are likely to continue indefinitely. Quarantine and massive chemical campaigns are only extremely expensive ways of buying time. We are faced, according to Dr. Elton, "with a life-and-death need not just to find new technological means of suppressing this plant or that animal"; instead we need the basic knowledge of animal populations and their relations to their surroundings that will "promote an even balance and damp down the explosive power of outbreaks and new invasions."

Much of the necessary knowledge is now available but we do not use it. We train ecologists in our universities and even employ them in our governmental agencies but we seldom take their advice. We allow the chemical death rain to fall as though there were no alternative, whereas in fact there are many, and our ingenuity could soon discover many more if given opportunity.

Have we fallen into a mesmerized state that makes us accept as inevitable that which is inferior or detrimental, as though having lost the will or the vision to demand that which is good? Such thinking, in the words of the ecologist Paul Shepard, "idealizes life with only its head out of water, inches above the limits of toleration of the corruption of its own environment. . . . Why should we tolerate a diet of weak poisons, a home in insipid surroundings, a circle of acquaintances who are not quite our enemies, the noise of motors with just enough relief to prevent insanity? Who would want to live in a world which is just not quite fatal?"

Yet such a world is pressed upon us. The crusade to create a chemically sterile, insect-free world seems to have engendered a fanatic zeal on the part of many specialists and most of the so-called control agencies. On every hand there is evidence that those engaged in spraying operations exercise a ruthless power. "The regulatory entomologists . . . function as prosecutor, judge and jury, tax assessor and collector and sheriff to enforce their own orders," said Connecticut entomologist Neely Turner. The most flagrant abuses go unchecked in both state and federal agencies.

It is not my contention that chemical insecticides must never be used. I do not contend that we have put poisonous and biologically potent chemicals indiscriminately into the hands of persons largely or wholly ignorant of their potentials for harm. We have subjected enormous numbers of people to contact with these poisons, without their consent and often without their knowledge. If the Bill of Rights contains no guarantee that a citizen shall be secure against lethal poisons distributed either by private individuals or by public officials, it is surely only because our forefathers, despite their considerable wisdom and foresight, could conceive of no such problem.

I contend, furthermore, that we have allowed these chemicals to be used with little or no advance investigation of their effect on soil, water, wildlife, and man himself. Future generations are unlikely to condone our lack of prudent concern for the integrity of the natural world that supports all life.

There is still very limited awareness of the nature of the threat. This is an era of specialists, each of whom sees his own problem and is unaware of or intolerant of the larger frame into which it fits. It is also an era dominated by industry, in which the right to make a dollar at whatever cost is seldom challenged. When the public protests, confronted with some obvious evidence of damaging results of pesticide applications, it is fed little tranquilizing pills of half truth. We urgently need an end to these false assurances, to the sugar coating of unpalatable facts. It is the public that is being asked to assume the risks that the insect controllers calculate. The public must decide whether it wishes to continue on the present road, and it can do so only when in full possession of the facts. In the words of Jean Rostand, "The obligation to endure gives us the right to know."

Where and why does Carson use value-laden rather than neutral language? Write a paragraph describing why we should or shouldn't (1) use pesticides to increase crop yields and (2) sacrifice forests to land developers. Try to persuade your audience to act.

Richard Selzer

Liver

When Richard Selzer turned 40 years old, he was a professor of surgery at Yale Medical School. He also ran a busy private practice. However, he wanted to write. He drew on his experience as a physician to write a fictional account of the plight of Jonah: "Since I know what the gastric mucosa looks like better than Biblical authors, I was able to describe Jonah's new home rather more authentically than had ever been done before." That article encouraged Selzer to write more. He later sold a horror story to Ellery Queen's *Mystery Magazine* and short stories to *The New American Review* and *Esquire*.

"Liver" is Selzer's first piece of nonfiction written for *Esquire*. It won the National Magazine Award.

What is the size of a pumpernickel, has the shape of Diana's helmet, and crouches like a thundercloud above its bellymates, turgid with nourishment? What has the industry of an insect, the regenerative powers of a starfish, yet is turned to a mass of fatty globules by a double martini (two ounces of alcohol)? It is . . . the liver, doted upon by the French, assaulted by the Irish, disdained by the Americans, and chopped up with egg, onion, and chicken fat by the Jews.

Weighing in at three to four pounds, about one fiftieth of the total body heft, the liver is the largest of the glands. It is divided into two great lobes, the right and left, and two small lobes, the caudate and the quadrate, spitefully named to vex medical students. In the strangely beautiful dynamism of embryology, the liver appears as a tree that grows out of the virgin land of the foregut in order to increase its metabolic and digestive function. Its spreading crown of tissue continues to draw nourishment from the blood vessels of the intestine. Legion are the functions of this workhorse, the most obvious of which is the manufacture and secretion of a pint of bile a day, without which golden liquor we could not digest so much as a single raisin; and therefore, contrary to the legend that the liver is an organ given to man for him to be bilious with, in its absence we should become rather more cantankerous and grouchy than we are.

I think it altogether unjust that as yet the liver has failed to catch the imagination of modern poet and painter as has the heart and more recently the brain. The heart is purest theatre, one is quick to concede, throbbing in its cage palpably as any nightingale. It quickens in response to the emotions. Let danger threaten, and the thrilling heart skips a beat or two and tightrope-walks arrhythmically before lurching back into the forceful thump of fight or flight. And all the while we feel it, hear it even—we, its stage and its audience.

One will grant the heart a modicum of history. Ancient man slew his enemy, then fell upon the corpse to cut out his heart, which he ate with gusto, for it was well understood that to devour the slain enemy's heart was to take upon oneself the strength, valor and skill of the vanquished. It was not the livers or brains or entrails of saints that were lifted from the body in sublimest autopsy, it was the heart, thus snipped and cradled into worshipful palms, then soaked in wine and herbs and set into silver reliquaries for the veneration of the faithful. It follows quite naturally that Love should choose such an organ for its bower. In the absence of Love, the canker gnaws it; when Love blooms therein, the heart dances and *tremor cordis* is upon one.

As for the brain, it is all mystery and memory and electricity. It is enough to know of its high-topping presence, a gray cloud, substantial only in the bony box of the skull and otherwise melting into a blob of ghost-colored paste that can be wiped up with a sponge. The very idea of it, teeming with a billion unrealized thoughts, countless circuits breaking and unbreaking, flashing tiny fires of idea on and off, is too much. One bows before the brain, fearing, struck dumb. Or almost dumb, saying pretty things like, "The brain is wider than the sky," or silly things like, "The brain secretes thought as the stomach gastric juice, the kidneys urine." Or this: "Rest, with nothing else, corrodes the brain," which is a damnable lie.

It is time to turn aside from our misplaced meditation on the privileged brain, the aristocratic heart. Let the proletariat arise. I give you . . . the liver! Let us celebrate that great maroon snail, whose smooth back nestles in the dome of the diaphragm, beneath the lattice of the rib cage, like some blind wise slave, crouching above its colleague viscera, secret, resourceful, instinctive. No wave of emotion sweeps it. Neither music nor mathematics gives it pause in its appointed tasks. Consider first its historical role.

Medicine, as is well known, is an offshoot of religion. The predecessor of the physician as healer was the priest as exorciser. It is a quite manageable leap for me from demons to germs as the source of disease. It is equally easy to slide from incantation to prescription. Different incantations for different diseases, I gather, and no less

mysterious to the ancient patients than the often mystic formulate of
one's family doctor. The mystery was and is part of it. Along with the
priest as exorciser was the priest as diviner, who was able to forestall
illness by his access to the wishes of the gods, a theory that has since
broadened into the field of preventive medicine. The most common
method of divination was the inspection of the liver of a sacrificial
animal, as is documented in cultures ranging from the Babylonian
through the Etruscan to the Greek and Roman. Why the liver as the
"organ of revelation par excellence"? Well, here it is:

In the beginning it was the liver that was regarded as the center of
vitality, the source of all mental and emotional activity, nay, the seat of
the soul itself. Quite naturally, the gods spoke therein. What a
gloriously hepatic age! A man could know when and how best to attack
his enemy, whether his amorous dallying would bring joy
unencumbered with disease, whether the small would be great, the
great laid low. All one had to do was to drag a sheep to the temple, flip
a drachma to an acolyte, and stand by while the priests slit open the
belly and read the markings of the liver, the position of the gallbladder,
the arrangement of the ducts and lobes. It was all there, in red and
yellow. This sort of thing went on for three thousand years and, one
might ask, what other practice has enjoyed such longevity? Even so
recent a personality as Julius Caesar learned of the bad vibes of the Ides
of March from an old liver lover, although that fellow used a goat
instead of a sheep, and a purist might well have been skeptical.
Incidentally, the reason the horse was never used for divination is that it
is difficult to lift onto altars, and also does not possess a gallbladder, a
fact of anatomy that has embarrassed and impoverished veterinarians
through the ages.

It was only with the separation of medicine from the apron strings
of religion and the rise of anatomy as a study in itself that the liver was
toppled from its central role and the heart was elevated to the chair of
emotions and intellect. The brain is even more recently in the money,
and still has not quite overcome the heart as the seat of the intellect, as
witness the quaint reference to learning something "by heart." Soon the
heart was added to the organs used for prophecy by the Greeks and
Romans, who then threw in the lungs, and finally, with an overdeveloped
sense of organic democracy, the intestines. Since the liver was no
longer *the* divine organ in the animal, out it transited along with
Ozymandias and other sic glorias, which decline so dispirited the
hepatoscopists that soon they gave up the whole damned rite and went
off to listen sulkily to Hippocrates rhapsodize, tastelessly I think, about
the brain and heart. As if that were not bad enough, Plato placed the
higher emotions, such as courage, squarely above the diaphragm, and
situated the baser appetites below, especially in the liver, where they
squat like furry beasts even today, as is indicated in the term

"lily-livered," or "choleric," or worse, "bilious." The assassination was complete. Still, there are memories, and the sense of history is a power and glory to all but the most swinish of men.

Today all that is left of the practice of divination is the unofficial cult of phrenology, in which character is interpreted from the bumps of the skull, and the science of palmistry, which is a rather ticklish business to get into, and seems to me merely a vulgar attempt to transfer divination to a more accessible part of the body. Nevertheless, my palmist always places great emphasis on the length and curve of the hand marking known as the "line of the liver."

The closest thing to liver worship still in business is the reverent sorrow with which the French regard their beloved *foie*. It is at once recalcitrant child and stern pater-familias. This national preoccupation is entirely misunderstood, and thus held in ill-deserved contempt by the rest of the world, which regards such hepatism as a form of mass hypochondria. In fact, it is wholly admirable after the shallow insouciance and hectic swilling of the Americans, for instance. The French understand the absolute fairness of life, that if you want to dance you have to pay the fiddler. Nothing in our country so binds the populace into a single suffering fraternity, any of whose members has but to raise his eyebrows and tap himself below the ribs to elicit a heartfelt moue of commiseration from a passing stranger. In the endless discussions of the relative merits of the various mineral waters, or whether the cure taken at Vichy or Montecatini effects a more enduring remission, class distinctions become as vague as mist. Princes of the Church, Communists at the barricade, légionnaires d'honneur, and chimney sweeps lock hearts and arms in the thrill of fellowship. Napoleon himself, wintering bitterly near Moscow in 1812, yearned not for Paris, or the Seine, or Versailles, or even for Josephine, but for Vichy and *the cure*. Obviously it is in the camaraderie of the liver and not in fragile treaties or grudging coexistence that the hope of the world lies—for ardently though one might wish to wash the brain of one's enemy, to bomb, bug or hijack him, never would one sink to the infliction of harm upon his liver.

That these same French viciously funnel great quantities of grain into the stomachs of their geese in order to fatten the livers for pâté de foie gras, I consider simply a regrettable transference of their own hepatic anxiety onto their poultry. It is as though by bringing on such barnyard *crises de foie* they are in some way exorcised of their own. Ah, the power of insight! To gain it is to forgive and to love.

Deplorable is the constipated English view, delivered with nasal sanctimony, that the Continental liver phobia is an old wives' tale. One has but to glance at any recent map of the British Empire to know the folly of this opinion. Equally regrettable is the *que sera, sera* attitude in the United States, where the last homage to the liver was paid by the faithful users of Carter's Little Liver Pills, until the federal government invaded even that small enclave of devotion by ordering the discontinuance of the word "Liver." Now, oh, God, it's just plain Carter's Little Pills and we are the poorer.

Man's romance with alcohol had its origins in the Neolithic Age or earlier, presumably from the accidental tasting by some curious fellow of, let's say, fermented honey, or mead as it is written in *Beowulf.* The attainment of the resultant euphoria has remained a continuous striving of the human race with the exception of the perverse era of Prohibition, which presumed to tear asunder that which Nature had joined in absolute harmony. Even in the Scriptures it is implied that Noah had a lot to drink, and Lot could not say Noah. From its first appearance on the planet, alcohol has never been absent from the scene. An early hieroglyphic of the Eighteenth Egyptian Dynasty has a woman calling out for eighteen bowls of wine. "Behold," she cries, "I love drunkenness."

The human body is perfectly suited for the ingestion of alcohol, and for its rapid utilization. In that sense we are not unlike alcohol lamps. Endless is our eagerness to devour alcohol. Witness the facts that it is absorbed not only from the intestine, as are all other foods, but directly from the stomach as well. It can be taken in by the lungs as an inhalant, and even by the rectum if given as an enema. Once incorporated into the body, it is to the liver that belongs the task of oxidizing the alcohol. But even the sturdiest liver can handle only a drop or two at a time, and the remainder swirls ceaselessly about in the bloodstream, is exhaled by the lungs and thus provides the state police with a crackerjack method of detecting and measuring the presence and amount of alcohol ingested. Along the way it bathes the brain with happiness, lifting the inhibitory cortex off the primal swamp of the id and permitting to surface all sorts of delicious urges such as the one to walk into people's houses wearing your wife's hat. Happily enough, the brain is not organically altered by alcohol unless taken in near-lethal amounts. The brain cells are not destroyed by it in any kind of moderate drinking, and if the alcohol is withdrawn from the diet, the brain rapidly awakens and resumes its function at the usual, if not normal, level. One must reckon, nevertheless, with the hangover, which retributive phenomenon is devised to make the drinker feel guilty. In fact, it is not more than a nightmarish echo of the state of inebriation brought on by excessive fatigue and the toxic effects of congeners, the natural products of the fermentation process that give distinction to the taste of the various forms of alcohol. In *The Adventures of Huckleberry Finn*, we find Huck awakening to "an awful scream. . . . There was pap looking wild, and skipping around every which way, and yelling about snakes. He said they was crawling up his legs; and then he would give a jump and scream, and say one had bit him on the cheek—but I couldn't see no snakes. He started and run round and round the cabin, hollering. 'Take him off. Take him off; he's biting me on the neck!" I never see a man look so wild in the eyes. Pretty soon he was all fagged out, and fell down panting; then he rolled over and over wonderful fast, kicking things every which way, and striking and grabbing at the air with his hands, and screaming, and saying there was devils ahold of him. He wore out by and by, and laid still awhile, moaning. Then he laid stiller, and didn't make a sound."

It was a French physician, quite naturally, who first described the disease known as cirrhosis of the liver, near the turn of the nineteenth century. His name, René Théophile Hyacinthe Laënnec. This fastidious gentleman was the very same whose aversion to applying his naked ear to the perfumed but unbathed bosoms of his patients inspired him to invent the stethoscope, which idea he plagiarized from a group of street urchins playing with rolled-up paper. The entire medical world continues to pay homage to Laënnec for his gift of space interpersonal. As if this were not enough, he permitted himself to be struck by the frequent appearance at autopsy of livers that were yellow, knobby and hard. This marvel he named cirrhosis, from the Greek word for tawny, *kirrhos*. The liver appears yellow because it is fatty, hard because it is scarred, and knobby because the regeneration of liver tissue between the scars produces little mounds or hillocks. It was suspected by Laënnec, and is known by all the rest of us today, that by far the most common cause of cirrhosis is the consumption of alcohol.

It is a matter for future anthropologists to ponder that the two favorite companions of business are Bottle and Board. More than one eminent literary agent, Wall Street broker, and vice-president have died testifying affection for them. Deep drinking and intrigue are part of all the noble professions. These, combined with the studius avoidance of exercise, have conspired to produce a whole race of voluptuaries who, by twos and threes from noon till three, sit at tables in dim restaurants, picking at their sideburns and destroying the furniture with their gigantic buttocks. These same men can be seen after five years of such indiscretion transformed into "lean and slippered pantaloons," with scanty hair that is but the gray garniture of premature senescence.

In the city of New York such is the torrent of spirituous flow as to make the clinking of ice cubes and the popping of corks a major source of noise pollution. It is as though it had been purported by the Surgeon General himself that the best means of maintaining human life from infancy to extreme old age were by the copious use of the Blood of the Grape. It might with equal credibility be put forth that tobacco smoke purifies the air from infectious malignancy by its fragrance, sweetens the breath, strengthens the brain and memory, and restores admiration to the sight.

Counting every man, woman and child in the United States, it is estimated that the average daily intake of calories is thirty-three hundred a person. In this all-inclusive group, one hundred sixty-five calories are ingested as alcohol. Pushing on, if one were to divide these Americans into I Do Drinks and I Don't Drinks, the I Do's take in five hundred calories from this same source. This alcohol is metabolized in the liver by a fiercely efficient enzyme called alcohol dehydrogenase, and transformed directly into energy, which would all be terribly nice were it not for the unjust fact that alcohol is poisonous to the liver, causing it to become loaded with fat. If enough is imbibed, and enough fat is deposited in the liver, this organ takes on the yellowish color noted by Laënnec. Still more booze, and the liver becomes heavy with fat, swelling so that it emerges from beneath the protective rib cage and bulges down into the vulnerable soft white underbelly. There it can be palpated by the examining fingers, and even seen protruding on the right side of the abdomen in some cases.

Even today, the progression from this fatty stage to the frank inflammation and scarring that are the hallmarks of cirrhosis is not well understood. Factors other than continued drinking pertain here. One of these is susceptibility. Jews, for instance, are not susceptible. One sees precious few cirrhotic Jews. It was formerly averred by somewhat chauvinistic Jewish hepatologists that Jews didn't get cirrhosis because they didn't drink much, what with their strong, dependable family ties, and their high motivation, and their absolute need to excel in order to survive. They didn't need to drink. But Jews are now among the most emancipated of drinkers and, with all the fervor of new converts, are causing such virtuosi as the Irish and the French to glance nervously over their shoulders. Still, the Jews do not get cirrhosis. This is not to say that they are not alcoholics. It has been reported by more than one visiting professor of medicine that noticeable segments of the population of Israel get and stay drunk for quite heroic periods of time. It is also reported that their livers remain enviably healthy.

Another measure of susceptibility is, brace yourselves, the absence of hair on the chest. In males, of course. Unpelted men of the sort idealized by bathing-suit and underwear manufacturers are sitting ducks for the onset of cirrhosis. All other things being equal, women, the marrying kind, would do well to turn aside from such vast expanses of naked chest skin and to cultivate a taste for the simian. It was formerly thought that cirrhotic men lost their chest hair. Not so. They never had any to begin with.

Lastly, it is said by some that climate is a factor: the closer to the equator, the more vulnerable the liver. Thus, a quantity of alcohol that scarcely ruffles the frozen current of a Norwegian's blood would scatter madness and fever into the brain of a Hindu.

There is a difference, I hasten to add, between imbibers of alcohol and alcoholics. Both develop fatty livers, true, but no one has shown conclusively that a fatty liver is the precursor of cirrhosis. One martini increases the fat content of the liver sufficiently so that it can be seen by the use of special stains under the microscope. In other words, a single martini increases the fat in a liver by one half percent of the weight of the organ, above a normal three percent. In the alcoholic this commonly reaches a death-defying twenty-five percent. But you don't have to be an alcoholic to get cirrhosis. Some quite modest drinkers get it. Nor does it matter the purity of the spirits consumed. Beer, wine, and whiskey equally offend, and he who would take comfort from the idea that he drinks only beer, or only wine, is to be treated with pity and contempt. One correlation that does hold water is the duration of time that one has been drinking. Cirrhosis is primarily a disease of the forties or fifties. Even here we cannot generalize, however, for great numbers of younger people are afflicted, and one patient within my ken was an eighteen-year-old girl whose voluminous liver could be felt abutting on her groin just eight months after she had retired to her room with a continuous supply of Thunderbird wine.

The state of nutrition is also a factor in the development of cirrhosis. It is no secret that boozers, the serious kind, stop eating, especially protein, either because they can't afford it—what with the cost of a bottle of bourbon these days—or because the sick liver just can't handle the metabolism of protein well, and the appetite is warned off.

The nitrogenous material of protein passes directly through the diseased liver and exerts a toxic effect on the brain. If one restricts protein in the diet of cirrhotics, the brain improves. A case in point is Sir Andrew Aguecheek of *Twelfth Night*, whose fervent wish to cut a dash was aborted by stupidity, cowardice, and social gaucherie. His eccentricity, emotional lability and restricted vocabulary were almost certainly due to the organic brain syndrome of liver disease due to intolerance of nitrogen. Sir Toby Belch assesses Sir Andrew rather highly. Still, Sir Toby cannot resist the clinical judgment that "For Andrew, if he were opened, and you find so much blood in his liver as will clog the foot of a flea, I'll eat the rest of the anatomy." Such as Sir Andrew Aguecheek are thrown into mental confusion, confabulation and even coma by no more than a single ounce of beef. Thus their medical nickname, "one meatballers."

In an analysis of the inhabitants of Chicago's Skid Row, it was observed that a customary diet consisted of alcohol in any form and jelly doughnuts. Yet in the cases of thirty-nine hundred such folk whose death certificates were signed out as cirrhosis, only ten percent were actually found to have the disease at autopsy. Thus it might be stated that alcoholics exceed cirrhotics by nine to one—or that only ten percent of alcoholics get cirrhosis.

What is clearly needed is a test to find out which are the ten percent that are going to get it, so that the rest of us can enjoy ourselves. At the moment I prefer to take comfort from the example of such valiant topers as Winston Churchill, who swallowed a fifth of whiskey a day all the while leading Great Britain in her finest hour, and went on to die in his nineties, still holding his fingers up like that. It is also true that if one, moved by some transcendental vision or goaded by ill-conceived guilt, abstains from further drinking, in short order all the excess fat departs from the liver and it once again regains its pristine color and size. In this way do spree drinkers inadvertently rest their livers and avoid the cirrhosis we slow but steadies risk. Thus something can be said for periodic abstinence, a wisdom one would hesitate to translate into other vices.

Before enumerating the signs and symptoms of cirrhosis, and thus running the risk of offending sensibility, it should be unequivocally affirmed that he is no gentleman, in fact a very milksop, of no bringing up, that will not drink. He is fit for no company, for it is a credit to have a strong brain and carry one's liquor well. Saith Pliny, "'Tis the greatest good of our tradesmen, their felicity, life and soul, their chiefest comfort, to be merry together in a tavern."

Envision, if you will, a house whose stones are living hexagonal tiles not unlike those forming the bathroom floors of first-class hotels. These are the hepatocytes, the cellular units of the liver. Under the microscope they have a singular uniformity, each as like unto its fellow as the antlers of a buck, and all fitted together with a lovely imprecision so as to form a maze of crooked hallways and oblong rooms. Coursing through this muralium of tissue are two arborizations of blood vessels, the one bringing food and toxins from the intestine, the other delivering oxygen from the heart and lungs. Winding in and among these networks is a system of canaliculi that puts to shame all the aqueductal glories of Greece and Rome. Through these sluice the rivers of bile, gathering strength and volume as the little ducts at the periphery meet others, going into ones of larger caliber, which in turn fuse, and so on until there are two large tubes emerging from the undersurface of the liver. Within this magic house are all the functions of the liver carried out. The food we eat is picked over, sorted out, and stored for future use in the cubicles of the granary. Starch is converted to glycogen, which is released in the form of energy as the need arises. Protein is broken down into its building blocks, the amino acids, later to be fashioned into more YOU, as old tissues die off and need to be replaced. Fats are stored until sent forth to provide warmth and comfort. Vitamins and antibodies are released into the bloodstream. Busy is the word for the liver. Deleterious substances ingested, inadvertently like DDT or intentionally like alcohol, are either changed into harmless components and excreted into the intestine, or stored in locked closets to be kept isolated from the rest of the body. Even old blood cells are pulverized and recycled. Such is the ole catfish liver snufflin' along at the bottom of the tank, sweepin', cleanin' up after the gouramis, his whiskery old face stirrin' up a cloud of rejectimenta, and takin' care of everything.

But there are limits. Along comes that thousandth literary lunch and—Pow! the dreaded wrecking ball of cirrhosis is unslung. The roofs and walls of the hallways, complaining under their burden of excess fat, groan and buckle. Inflammation sets in, and whole roomfuls of liver cells implode and die, and in their place comes the scarring that twists and distorts the channels, pulling them into impossible angulation. Avalanches block the flow of bile and heavy tangles of fiber impede the absorption and secretion. This happens not just in one spot but all over, until the gigantic architecture is a mass of sores and wounds, the old ones scarring over as new ones break down.

The obstructed bile, no longer able to flow down to the gut, backs up into the bloodstream to light up the skin and eyes with the sickly lamp of jaundice. The stool turns toothpaste white in commiseration, the urine dark as wine. The belly swells with gallons of fluid that weep from the surface of the liver, no less than the tears of a loyal servant so capriciously victimized. The carnage spreads. The entire body is discommoded. The blood fails to clot, the palms of the hands turn mysteriously red, and spidery blood vessels leap and crawl on the skin of the face and neck. Male breasts enlarge, and even the proud testicles turn soft and atrophy. In a short while impotence develops, an irreversible form of impotence which may well prod the invalid into more and more drinking.

Scared? Better have a drink. You look a little pale. In any case there is no need to be all that glum. Especially if you know something that I know. Remember Prometheus? That poor devil who was chained to a rock, and had his liver pecked out each day by a vulture? Well, he was a classical example of the regeneration of tissue, for every night his liver grew back to be ready for the dreaded diurnal feast. And so will yours grow back, regenerate, reappear, regain all of its old efficiency and know-how. All it requires is quitting the booze, now and then. The ever-grateful, forgiving liver will respond joyously with a multitude of mitoses and cell divisions that will replace the sick tissues with spanking new nodules and lobules of functioning cells. This rejuvenation is carried on with the speed and alacrity of a starfish growing a new ray from the stump of the old. New channels are opened up, old one dredged out, walls are straightened and roofs shored up. Soon the big house is humming with activity, and all those terrible things I told you happen go away—all except that impotence thing. Well, you didn't expect to get away scot-free, did you?

And here's something to tuck away and think about whenever you want to feel good. Sixty percent of all cirrhotics who stop drinking will be alive and well five years later. How unlike the lofty brain which has no power of regeneration at all. Once a brain cell dies, you are forever one shy.

Good old liver!

How would you describe Selzer's writing? Personal? Poetic? Detailed? Give examples to support your answer.

Exercises

1. Discuss, support, or refute the ideas contained in these quotations:

   Every individual alive today, the highest as well as the lowest, is derived in an unbroken line from the first and lowest forms. —August Frederick Leopold Weismann

   If you cannot—in the long run—tell everyone what you have been doing, your doing has been worthless. —Erwin Schrödinger

   There are passages in every novel whose first writing is the last. But it's the joint and cement between these passages that take a great deal of rewriting. . . . Each sentence is a skeleton accompanied by enormous activity of rejection. —Thornton Wilder

   One hundred trout are needed to support one man for a year. The trout, in turn, must consume 90,000 frogs, that must consume 27 million grasshoppers that live off of 1,000 tons of grass. —G. Tyler Miller, Jr.

   In science the important thing is to modify and change one's ideas as science advances. —Herbert Spencer

   Science is the father of knowledge, but opinion breeds ignorance. —Hippocrates

   Science has given to this generation the means of unlimited disaster or of unlimited progress. There will remain the greater task of directing knowledge lastingly towards the purpose of peace and human good. —Winston Churchill

2. Simplify this sentence taken from an anatomy textbook:

   The lateral surface, lateral to the anterior border, is anterior, lateral, and also posterior above, for it extends from the anterior to the posterior border of the radial tuberosity, but is largely lateral below, and the posterior surface, narrow and mostly medial above, expands and is truly posterior.

3. Educators are notorious for their poor writing. Revise this sentence taken from a book:

   Operationally, teaching effectiveness is measured by assessing the levels of agreement between the perceptions of instructors and students on the rated ability of specific instructional behavior attributes which were employed during course instruction.

4. Simplify these sentences and paragraphs:

Pollutants exist in virtually all sectors of the environment.

I only did research for two years.

It is generally desirable to communicate your thoughts and ideas in a directly forthright manner and style. Toning down your point and tiptoeing around it may, in a large number of circumstances, tempt the reader to tune out and allow her mind to wander.

They are without any nutrients whatsoever.

Some serious problems can be prevented by means of the use of effective writing.

There is no doubt but that the animals expired.

Basically, it would not be reasonable to assume that in the foreseeable future our very unique but nonetheless inefficient experiments will have to be terminated.

In no case whatsoever did any of the laboratory mice develop lesions.

The consensus of opinion is that the experiment will not be completely finished until next week.

There were on the order of about 200 apes.

Your method of mixing together the chemicals is very unique.

It is apparent that they want to hire a male or female with scientific experience in the field of biology.

Our current belief is that auxin and calcium control root gravitropism.

We want results that are exactly alike.

In four instances the reagents were red.

The nose is part of the respiratory mechanism that plays an important part in the normal function of breathing.

The degree of importance in the level of accuracy depends upon the individual situations.

It is necessary to have complete documentation of your research.

Pursuant to the regulations promulgated as of this date by this author, endeavor to employ uncomplicated words in writing.

The aquarium was two meters in its length-wise dimension.

Watson and Crick's ideas were radically new.

There is the possibility that they may finish the experiment before us.

I do not have sufficient knowledge of the problem to make a proposal regarding a new analysis.

Be sure to communicate your comments to your teacher.

In order to evaluate the potential significance of certain molecular parameters at the ultracellular and subcellular levels, and so throw some light on the conceivable role of structural configuration in spatial relationships of intracellular and intercellular molecular structures, a comprehensive and integrated approach to the problem of cellular structure and function has been developed. The results, which are in a preliminary stage, are discussed here in quite a bit of detail because of their potential implications in mechanisms of cellular structure and function in a larger context.

Your instructor will not accept a report turned in late.

It takes a longer period of time to check these solutions than to create them from scratch in the first place.

In the event that the chemicals need to be stored prior to their utilization, it is preferred that they be stored in a warehouse before they are unpacked.

I have been trying to elucidate the solution to this interesting problem for a period on the order of about a decade.

Because of the unfortunate fact that it happened to be snowing at that particular point in time, we came to the conclusion that we would refrain from beginning our sampling of the Marlin Prairie.

We are working toward a unanimous situation regarding our data.

After being ingested in the diet, zinc is absorbed from the digestive track, partly arresting the chance of poisoning, but, nevertheless, in effect, its corrosive characteristics and traits often and frequently produce gastro-intestinal irritation that leads to nauseousness as well as an inflammation of the stomach and gastric necrosis.

An increased appetite was manifested by all of the dogs at that point in time.

Experiments are currently in progress to assess the possibility of using the new spectrophotometer.

It consists essentially of two parts.

They all arrived at a consensus of opinion in regard to the basic principle that the continuing utilization of the antiquated methodology in the lab is out-dated and generally inefficient.

During the course of his or her education, it is absolutely essential that a student secure the acquisition of the basic elements and foundation of effective writing.

As a result of the fact that our biology students could not be present at our last scheduled lab meeting time, the liberty was taken to cancel that gathering. Therefore, it is planned to consider the identical research items at the newly scheduled gathering of the students that will gather Tuesday. This gathering will gather in accordance with the same agenda, repeated herein for your convenient reference.

Asbestos has been known to be implicated as a causal factor regarding cancer of the chest has obvious beneficial utilizations.

The stomach is oval in shape.

A group of people who contributed exemplary service to the diffusion of knowledge of biology were the early cell biologists. It goes without saying that without them biological research would have been slowed down a lot. It was only after the discovery of the microscope that such basic and fundamental biological sciences such as microbiology and histology were started.

The general feeling of the meeting was that within the basic framework of the experiments a great amount was definitely accomplished and cognitively learned by all participants, and the very unique system has achieved virtually all of our objectives with regard to our experiments.

In order to evaluate the potential significance of certain molecular parameters at the subcellular level, and to achieve an understanding of the possible role of configuration and structure in spatial relationships of intracellular macromolecules, a comprehensive, integrated, and multifaceted approach to the ongoing problem of intercellular signal transduction has been devised, developed, and implemented. The results, which are currently in a preliminary stage at this point, are discussed here in a rather large amount of detail because of their possible implication regarding the mechanisms of intercellular communication in a wider and broader sphere.

At that point in time we were using a Zeiss microscope in the laboratory.

There is a distinct possibility that they will again repeat our experiment.

We need a recording system that does not make any noise whatsoever.

An increased appetite was manifested by all of the rats.

The answer is in the negative.

She pursued the tasks with great diligence.

To effect a proper utilization of time, ensure that preparation of the growth medium is absolutely complete before lab is initiated.

A variety of stimulatory chemical agents, irregardless of their chemical structure, composition, and configuration, are characterized by their uncanny ability to influence the production and synthesis of protein as a prerequisite for later secondary biological events characteristic of the particular target organ.

The basic purpose of this accumulation of writings is to first assist writers in the writing of biological papers in an acceptable way that helps the writer learn biology and secondly to possibly suggest the inclusion of greater uniformity in the overall approach to the preparation of biological papers for publication. Nevertheless, it is clear to this writer that specialized aspects of these topics may call for special methods of presentation and that these writings should be read in conjunction with other writings such as those in journals and books.

5. Simplify these words and phrases:

assert	a second point is that
transmit	demonstrate
draw to your attention	in view of the fact that
in a number of cases	in the not-too-distant future
initiate	commence
terminate	seem to suggest
aggregate	converse
have a tendency to	that point in time
as of this date	endeavor
attempt	on two separate occasions
anomalous	utilize
make use of	in most cases
in the majority of cases	is desirous of
had to occasion to be	proved to be
provided that	by the time
at such time as	the question as to whether
during the time that	if it should happen that
until such time as	final outcome
for the purpose of	at the membrane level
general rule	new innovation
positive identification	proposed plan
single unit	component parts
doctorate degree	weather conditions
brief in duration	classify into groups
estimated at about	filled to capacity
last of all	rectangular in shape
new all-time record	might possibly
mutually agreeable	assemble together
cancel out	attached hereto

connect up	enclosed herein
follow after	made out of
join together	at a later date
both together	have need for

6. Write a short essay describing how this *The Far Side* cartoon relates to biology:

Figure 3–1
The Far Side. Copyright 1988. UNIVERSAL PRESS SYNDICATE. Reprinted with permission.

C H A P T E R   F O U R

Precision

---

and Clarity

A clear statement is the strongest argument.
   —English proverb

For what good science tries to eliminate, good art
seeks to provoke—mystery, which is lethal to the
one, and vital to the other.
   —John Fowles

It is with words that we do our reasoning, and
writing is the expression of our thinking
. . . Words and phrases that do not have an exact
meaning are to be avoided because once one has
given a name to something, one immediately has a
feeling that the position has been clarified, whereas
often the contrary is true.
   —W.I.B. Beveridge, *The Art of Scientific
      Investigation*

Have something to say, and say it as clearly as you
can. That is the only secret of style.
   —Matthew Arnold

Some books about scientific writing would have biologists merely count syllables and words to satisfy the writing cliché "be brief." Although substituting simple words for complex, large words almost always improves communication, it's only a good first step. Concentrating only on simple words will not produce effective writing, because counting letters and syllables doesn't explain what makes a sentence awkward, much less tell you how to make it into a clear one.

In scientific writing, precision is the most important goal of language. If your writing does not communicate *exactly* what you think or did, then you have changed your ideas or research. Furthermore, most readers become annoyed with vague writing. Fairly or not, and often without realizing what they're doing, readers conclude that an imprecise writer is lazy and that your ideas aren't worth much of their attention. Vague writing, which discredits your work, results from not only a lack of grace but also a lack of clarity. I argue that these traits are not separable.

When you choose a precise word, trust it to do its job. If you've picked the right word, adding bankrupt words such as *quite* only impedes communication and defeats your purpose. Similarly, adding redundant modifiers such as *very* provides no compensation when you've chosen the wrong word. However, don't be overly precise; for example, don't spend five pages describing how an electron microscope works if all that you need is an electron micrograph.

Precise and concise writing says exactly what you mean and requires that you choose the right words. This means making every word tell.

Make Every Word Tell

Words, when well chosen, have so great a force in
them that a description often gives us more lively
ideas than the sight of the things themselves.
   —Joseph Addison

Vigorous writing is concise. A sentence should
contain no unnecessary words, a paragraph no
unnecessary sentences, for the same reason that a
drawing should have no unnecessary lines and a
machine no unnecessary parts. This requires not that
the writer make all his sentences short, or that he
avoid all detail and treat his subjects only in an
outline, but that every word tell.
   —William Strunk

Words may be either servants or masters. If the
former they may safely guide us in the way of truth.
If the latter they intoxicate the brain and lead into
swamps of thought where there is no solid footing.
   —Bishop George Horne

*Memorable sentences are memorable on account of
some single irradiating word.*
   —Alexander Smith, On the Writing of Essays

Precise writing is concise; every word tells. Precision is based largely on word choice. As Mark Twain said, "Use the right word, not its second cousin. The difference between the right word and the almost right word is the difference between 'lightning' and 'lightning bug.'" The *United Press International Stylebook* says it better: "A burro is an ass. A burrow is a hole in the ground. As a [writer] you are expected to know the difference."

Archie Bunker and Yogi Berra became famous partly because they often chose the almost right word instead of *the* right word. Choosing the right word is often difficult because many words have several meanings. For example, the 500 most-used words in English have a total of 14,000 meanings—an average of 28 meanings per word.

There are many ways to make every word tell. Having deleted the clutter from your writing is a good start, but you must do more. Specifically, you must know the meaning of every word you use. Don't settle for an *almost* right word in place of *the* right word. If you do, you'll confuse or annoy readers. You might also end up with sentences like these:

Darwin was a child progeny who wrote *Organ of the Species.*

Sir Francis Drake circumcised the world with a 100-foot clipper.

Three kinds of blood vessels are arteries, vanes, and caterpillars.

Although the key words in these sentences are almost correct, they're not *the* correct words. Consequently, the writers appear either careless, confused, or poorly educated. Save yourself embarrassment by buying and using a dictionary and a thesaurus.

Scientific writing must be accurate, precise, and clear, all of which depend on word choice.[1] While precision depends on choosing the right word, clarity often depends on not choosing the wrong word. This requires that you know the meaning and purpose of every word that you use. If you don't know the meaning of words that you use, you risk looking foolish. For example, the authors of an introductory biology book confused the words *mole* and *molecule* in their discussion of cellular respiration, claiming that 36 ATP is the "approximate maximal ATP yield from the complete respiration of 1 mole of glucose."[2] The authors go on to state that each of the 20 to 30 trillion cells in our bodies cleaves 1 to 2 billion ATP per minute. Simple arithmetic shows the inconsistency:

---

[1]Garbled writing has plagued people for ages. According to the Bible, when God wanted to stop people from building the Tower of Babel, God did not zap them with a thunderbolt. Rather, God said ". . . let us go down, and there confound their language, that they may not understand one another's speech." Apparently God could think of no better way to foil the project than to garble their words.

[2]Wessells, Norman K., and Janet L. Hopson, 1988. *Biology*, 1st edition. New York: Random House, pp. 165–166.

If each cell cleaves 1.5 billion ATP per minute, then each cell needs 41,700,000 (1.5 billion ATP/36 ATP per mole of glucose) moles of glucose per minute.

This means that each cell needs 7,506,000,000 (41,700,000 moles of glucose × 180 grams per mole) grams of glucose per minute.

To supply this much glucose to each of the 25 trillion cells in our body, we would each need to eat about 187,650,000,000,000,000,000,000,000 g (7,506,000,000 g of glucose per minute per cell × 25,000,000,000,000 cells) of glucose per minute. That's equal to a staggering 2.7×10^{23} kg of glucose per day, an amount equal to half the mass of the planet Earth. Even my huge cousin doesn't eat that much when, to use his colorful words, he roars through the kitchen "hungrier than the gang on Noah's ark."

Know the meanings of all the words that you use. Don't use *mole* when you mean *molecule*, *less* when you mean *fewer*, *allude* when you mean *refer*, *imply* when you mean *infer*, *weight* when you mean *mass*, *disinterested* when you mean *uninterested*, and *comprise* when you mean *compose*.

VAGUE: Our sampling at Walnut Creek was delayed for a time because of unfavorable weather conditions.

SPECIFIC: Our sampling at Walnut Creek was delayed for a week because of snowstorms.

Appendix 2 provides a list of words frequently misused by biologists and other scientists. Use these words carefully and, when in doubt, consult a dictionary or usage guide such as the *Harper Dictionary of Contemporary Usage*.

Avoid Doublespeak

Ready-made phrases are the prefabricated strips of words . . . that come crowding in when you do not want to take the trouble to think through what you are saying. . . . They will construct your sentences for you—even think your thoughts for you, to a certain extent—and at need they will perform the important service of partially concealing your meaning even from yourself.
    —George Orwell

It isn't that jargon is noxious in itself, it's that, like crabgrass, the dratted stuff keeps rooting where it doesn't belong.
    —Bruce O. Boston

*You must learn to talk clearly. The jargon of scientific
terminology which rolls off your tongue is mental
garbage.*
    —Martin H. Fischer

*The trouble with new or different words to express
the old ideas is that the words can give a false sense
of progress and disguise history.*
    —Thelma Ingles

A good catchword can obscure analysis for fifty years.
    —Wendell L. Willkie

*Anyone who uses the words "parameter" and
"interface" should never be invited to a dinner party.*
    —Dick Cavett

Whenever ideas fail, men invent words.
    —Martin H. Fischer

Doublespeak is a type of writing that pretends to communicate but really doesn't. Rather, it deliberately retreats from what is tangible, real, and common. For example, on March 24, 1989, the Exxon *Valdez* hit a reef and spilled 11 million gallons of North Slope crude oil into Prince William Sound. David Parish, a spokesperson for Exxon, said that Exxon did not expect major environmental damage from the spill and boldly claimed that Exxon was responsible for cleaning up the spill. However, when Exxon realized the magnitude of the spill—more than 1,000 miles of coastline had been contaminated—it used doublespeak to change its story: Exxon no longer said it would clean up the spill but instead said it would make the area "environmentally stable." Since no one knows exactly what "environmentally stable" means, Exxon let itself off the hook. Welcome to the smoke-and-mirrors world of doublespeak.

Doublespeak is all around us and is used to deceive, distort, confuse, or mislead us.[3] For example, potholes in streets become "pavement deficiencies";

[3]Andy Rooney knew the purposes of scientific jargon when he wrote, "Much of the language of science, medicine, government and law is nonsense designed to exclude outsiders so they won't discover that, basically, the specialists' work is not so complex as it seems." Similarly, George Orwell, in his famous essay "Politics and the English Language," wrote, "The great enemy of clear language is insincerity. When there is a gap between one's real and one's declared aims, one turns as it were instinctively to long words and exhausted idioms, like a cuttlefish squirting out ink." In the nightmarish world of his novel *1984*, Orwell described how a totalitarian government—"Big Brother"—used a form of doublespeak called Newspeak to create a reality desired by the government. Newspeak epitomized the deceptive power of doublespeak: It diminished thought so much that it allowed people to believe in conflicting ideas. Indeed, the height of doublespeak (and the doublethink that it caused) was the famous phrase, "War is peace." Lest you think that such doublespeak-induced contradictions are restricted only to Orwell's novel, recall that Secretary of State Alexander Haig said several times that a continued weapons build-up by the United States is "absolutely essential to our hopes for meaningful arms reduction." Or recall that Senator Orrin Hatch said, "Capital punishment is our society's recognition of the sanctity of human life." Such contradictions are doublespeak.

poor people become "fiscal underachievers"; medical malpractice becomes "diagnostic misadventure"; kickbacks become "rebates"; smelling becomes "organoleptic analysis," guards in department stores become "loss prevention specialists"; and gas station attendants become "petroleum transfer engineers." Airlines speak of "water landings" instead of crashes at sea, and the CIA subsidizes "health alteration committees," not assassination squads. Companies don't report losses; they speak of "negative growth," which is analogous to talking about the heat in an ice cube. To paraphrase a famous quote, "Nothing is certain in life except negative patient care outcome and governmental revenue enhancements."

Scientists often chuckle when they hear doublespeak. Words and phrases such as "pseudopseudohypoparathyroidism" and "nonrheumatoid rheumatoid nodules" are whimsical, and few are fooled by ludicrous phrases such as "oral hygiene appliance" instead of toothbrush or "personal time control center" instead of wristwatch. However, biologists and other scientists often use doublespeak. This doublespeak appears as jargon, euphemisms, gobbledygook, and inflated language.[4]

Jargon   The word *jargon* once referred to the unintelligible chatter of birds. Today, jargon is specialized "code" language, usually used to impress, not communicate, that is understandable by only a few people. Jargon usually arrives in clusters:

> It is believed that with the parameters that have been imposed by your administrators, a viable research program may be hard to evolve. Net net: If our program is to impact students to the optimum, meaningful interface with your management may be necessitated.

This kind of writing is long-winded and heavy-handed, and is what E.B. White calls "the language of mutilation" because it mutilates your meaning. When you eliminate the jargon, the message becomes more obvious:

> We believe that the limits set by your administrators may prevent an effective research program. If we want to involve our students, we need to talk with your administrators.

Jargon insulates members of a group from the outside world and excludes nonmembers from the group. Jargon is often pretentious, obscure, and esoteric language used to imply profundity, prestige, or authority. It makes the simple seem complex, the ordinary seem profound, and the obvious seem insightful. For example, you can unclog your drains with a "hydro blastforce cup" (that is, a plunger), and what was once a vacuum cleaner is now Hoover's "Dimension

---

[4]The National Council of Teachers of English (NCTE) annually awards a Doublespeak Award to officials who are "grossly deceptive, evasive, euphemistic, confusing or self-contradictory." Winners of the 1990 Doublespeak Award included the Mobil Corporation for claiming that one of its trash bags is "photodegradable" even when the bag is buried in a landfill. For more information about the NCTE and its journal, *Quarterly Review of Doublespeak*, write to NCTE, 1111 Kenyon Road, Urbana, IL 61801.

1000 Electronic Cleaning Machine with Quadraflex Agitator." Physicians say "cholelithiasis" instead of "gallstones," and "deglutition" instead of "swallowing." Similarly, we'd probably react differently if a physician said we were being treated for "pyrexia" and "cephalalgia" than if told we were being given something to relieve our "fever" and "headache." However, we might be willing to pay more to get rid of "pyrexia" and "cephalalgia."

Scientists also use unnecessary jargon. For example, some refer to a "fused silicate container" instead of a glass beaker, and a crack in a test tube becomes a "structural discontinuity." Similarly, biologists at the Environmental Protection Agency write about "poorly buffered precipitation" and "atmospheric deposition of anthropogenetically derived acid substances," when what they really mean is acid rain. Physicians say "utilization of recently introduced therapeutic modalities" instead of "use of new treatments," and "expectorated a hemorrhagic production" instead of "spit out blood." As intended, such writing often isolates you from others. For example, behavioral scientists, instead of writing that "modern cities are hard to live in" might write, "urban existence in the perpendicular declivities of a stratocosmopolis or megalopolis. . . ." Such awful writing prompted writer-historian Barbara Tuchman to give an example of the influence of jargon:

> Let us beware the plight of our colleagues, the behavioral scientists, who by use of proliferating jargon have painted themselves into a corner—or isolation ward—of unintelligibility. They know what they mean, but no one else does.

If writing only for people with a background similar to yours, use technical terms. However, when writing to a more general audience, especially decision-makers with limited experience in your field, avoid jargon and use the simplest and most accurate words to express your ideas.

Euphemisms A euphemism is a positive or inoffensive word or phrase used to avoid a harsh or unpleasant reality. People use euphemisms to divert attention from a topic and to avoid communicating effectively.[5] For example, Adolph Hitler referred to murders as "the final solution," and in 1977 the Pentagon tried to obtain funds for a neutron bomb by calling it a "radiation enhancement device." Similarly, the U.S. State Department no longer refers to killings in reports about human rights. Rather, it writes about the "unlawful or arbitrary deprivation of life."

Euphemisms are always used to hide something. For example, dictators have used phrases such as "pacification," "terminating," and "protecting the peace"

---

[5]Several years ago the Justice Department filed a law-suit against the accounting firm Ernst and Whinney for intentionally using "false, misleading, and deceptive" language on a client's tax form. To qualify their client for tax credits, the accountants referred to windows as "decorative features," a fire alarm as a "combustion enunciator," doors as "movable partitions," 92 cubic yards of topsoil as a "planter," and a refrigerated warehouse as a "freezer." In their defense, Ernst and Whinney claimed that they were only trying "to put their client's best foot forward."

because they know these phrases do not create an image of bound, blindfolded, kneeling prisoners with pistols held to their heads. Similarly, consider all of our euphemistic fig leaves for the word "die": *croak, take the last count, pass away, pass on, pass to one's reward, go beyond, depart this world, expire, come to an untimely end, perish, be taken, give up the ghost, depart this life, kick the bucket,* and *buy the farm.* One biologist even described the death of a laboratory rat by saying that the rat "lost its integrity." Unless religious dictates prevent you from doing so, use the simpler and more dignified word *die.*

Biologists also use euphemisms to deceive readers. For example, biologists seldom write that they kill animals. Rather, they write that animals were "sacrificed," as if the animals had been part of some arcane religious ceremony. Similarly, biologists write about "negative patient care outcome" and "failure to fulfill his wellness potential" when all they mean is that the patient died and "national species of special emphasis" when they discuss endangered species that we can kill legally.

Avoid euphemisms. Choose simple, direct, more dignified statements.

Gobbledygook   During World War II, Congressman Maury Maverick of Texas attended a meeting where the chairperson spoke at length about "maladjustments co-extensive with problem areas . . . alternative but nevertheless meaningful minimae . . . utilization of factors in which a dynamic democracy can be channelized into both quantitative and qualitative phrases." Maverick censured the language by calling it *gobbledygook,* a term that would last far beyond the war.

Today, gobbledygook is synonymous with "bureaucratese," a writing style that masquerades as serious prose by overwhelming readers with big words, phrases, and sentences. Gobbledygook flourishes in universities and governmental agencies — an observation obvious to Lewis Thomas when he facetiously likened Agricultural Experiment Stations to governmental agencies "devoted to the breeding of new words."[6] For example, consider this statement by Alan Greenspan, then chairperson of President Nixon's Council of Economic Advisors: "It is a tricky problem to find the particular calibration in timing that would be appropriate to stem the acceleration in risk premiums created by falling incomes without prematurely aborting the decline in the inflation-generated risk premiums." *Say what?* The U.S. Department of Health, Education, and Welfare requested $23,000 for "the evaluation and parameterization of stability and safety performance characteristics of two- and three-wheeled vehicular toys for riding." In other words, they wanted to study how children fall off of bicycles

---

[6]Not all governmental administrators endorse the poor writing typical of bureaucrats. For example, soon after becoming the U.S. Secretary of Commerce, Malcolm Baldrige banned these words and phrases from the department's word-processors: *parameter, I would hope, I would like to express my appreciation, As I am sure you know, As you are aware, At the present time, bottom line, enclosed herewith, finalize, hopefully, institutionalize, It is my intention, maximize, Needless to say, new initiatives, prioritize, to impact, optimize, contingent upon, effectuated,* and *utilize.* When someone would type these words and phrases, the screen would flash, "Don't use this word."

and tricycles. A report issued by the U.S. Department of Education claims that "feediness is the shared information between toputness, where toputness is at a time just prior to the inputness." And consider this doozie: "The Arizona Career Ladder Research and Evaluation Project summated matrices depicting positive anecdotes related to interrelated organizational focus." As best I can tell, the translation of this gobbledygook is, "How a teacher's salary affects student morale and performance." Here's one from the medical literature: "In my opinion, and in that of others, only a massive, incessant, dynamic, multimedian educational campaign toward nationwide arousal of motivation and action, based on today's scientific knowledge, can offer any prospect for a gradual reduction of this country's unparalleled, unnecessary and uncontrolled premature cardiac mass mortality." Those 47 jawbreakers translate into this: "Only widespread health education will reduce the high rate of premature death caused by heart attack." Even writing teachers have joined the fray by claiming that students write better if they're taught "concretization of goals, procedural facilitation, and modeling planning." This kind of writing would overwhelm even the most sophisticated "shit-detector."

Inflated Language People use inflated language to change everyday observations into things that appear rare and important. Inflated language damages science because its unwieldy style reinforces the mystique of scientific writing being impenetrable. Inflated language is often used by scientists who fear that the substance of their idea will evaporate if they use simple, clear language. Unable to describe their ideas effectively, these writers retreat behind a cloud of ridiculous words and phrases. Here's how Nobel laureate Richard Feynman described the use of inflated language by NASA officials while investigating the *Challenger* accident:

> At any rate, the engineers all leaped forward. They got all excited and began to describe the problem to me. I'm sure they were delighted, because technical people love to discuss technical problems with technical people. . . .
> They kept referring to the problem by some complicated name—a "pressure-induced vorticity oscillatory wawa," or something.
> I said, "Oh, you mean a whistle."

Feynman, one of the most brilliant physicists of this era, was a passionate advocate of laypeople learning science. Consequently, he opposed and often chided scientist's use of doublespeak. Here's one of his descriptions of another official at NASA:

> Mr. Mulloy explains how the seal is supposed to work—in the usual NASA way: he uses funny words and acronyms, and it's hard for anybody else to understand.

Inflated language is often funny. For example, Pentagon writers call a toothpick a "wooden interdental stimulator" and a pencil a "portable hand-held communications inscriber." Others use inflated language to transform car mechanics into "automobile internists," elevator operators into members of the "vertical transportation corps," a black eye into a "circumorbital haematoma," food into "comestibles," used cars into "pre-owned vehicles" or "experienced automobiles," and undertakers into "perpetual rest consultants" who sell "underground condominiums" rather than cemetery lots. Other uses of inflated language aren't so easily deciphered. For example, when Chrysler "initiated a career alternative enhancement program," what it really did was lay off 5,000 workers. And never admit to having lousy handwriting; rather, just say you suffer from "restricted graphomotorial representation."

Although scientists and others scoff at such examples of doublespeak, they're often guilty of using it in their writing—not to reinvent the wheel, but merely to paint it another color. For example, many biologists administer "chemotherapeutic agents" rather than drugs, study a "three-dimensional biopolymer composite" rather than wood, and report on studies of "hematophagous arthropod vectors" rather than fleas. If you don't work for the Pentagon, follow Mark Twain's advice: "I never use the word *metropolis* because I'm always paid the same amount of money to write *city*."

Scientists often use doublespeak to make ideas that they think are too simple seem more impressive, much as some people use pompous, difficult-to-understand language to protect what they have from others who want it. Doublespeak keeps knowledge from others by hiding information behind language so impenetrable that only a select few can find it. Since science is a search for truth, "scholarly" writing that includes doublespeak hides the truth and does not advance science.

At its worst, doublespeak limits thought and encourages readers to believe in opposing ideas. At its least offensive, doublespeak is inflated language that tries to give importance to common or trivial aspects of science. It is ineffective because it inhibits communication and calls attention to the writing, not to the message. Eliminating doublespeak is especially important for biologists and other scientists because scientific writing is not a matter of having subject and verb agree but of having words and facts agree. Since doublespeak deceives or distorts the truth, it has no place in scientific writing. Avoid doublespeak by thinking straight and saying exactly what you mean.

Avoid Stacked Modifiers

We use modifiers such as adjectives and adverbs to describe our subjects and actions. Although effective writers use modifiers sparingly, such words are occasionally useful. However, when writers stack two or more modifiers in front of a subject or verb, readers have trouble deciding which word the first modifier is modifying. For example, consider this sentence: "The animals must be given more nourishing food." This is confusing. Does the author mean more food that's nourishing or the same amount of more nourishing food? In the phrase "normal dog hearts," what's normal—the hearts or the dogs? Other examples that have appeared in papers include "lizard ovary winter lipid level changes," "ankle joint angle measurement," "constant pressure heat capacity temperature maxima," "silica gel coated glass fiber paper chromatography," and "average resting and after arginine hydrochloride infusion plasma growth hormone concentration." These phrases are graceless and usually confuse readers. Cramming so much information together, without regard to style or audience, produces not an idea, but an indigestible lump.

Stacked modifiers are easily spotted and corrected. Simply count the number of modifiers preceding each noun in a sentence: If there is more than one modifier, rewrite the sentence to avoid the confusion. Although you may lose brevity by making these changes, you'll improve clarity.

Avoid Dangling Modifiers

When a sentence begins with a participial phrase, be sure that the phrase modifies the subject of the sentence. Otherwise, the phrase dangles in front of the wrong word as though it modifies that word instead of the one it should. Dangling modifiers often confuse readers:

> She found nine fish in the water, which was better than usual.
> (What was better than usual, the water or the number of fish?)

> After killing the rat, the diet was tested.
> (Did the diet kill the rat, or did someone kill the rat and then test the diet?)

Other times they produce sentences that are more amusing and embarrassing than confusing. For example, consider these gems:

> As a baboon who grew up in the wild, I realized that Tamba had strong sexual needs.

> When dipped in a weak acid, you can see the differential growth of roots.

> Mangy and needing a bath, I let the dog into the lab.

> Being old and dog-eared, I was able to buy the biology book for only $1.

> After soaking in sulfuric acid, I rinsed the seeds.

> Wondering what to do next, the incubator exploded.

> When 11 years old, my grandmother died.

> Quickly summoning an ambulance, the corpse lay motionless.

> Running through the lab, his hat blew off.

> After germination, I will grow the plants in the incubator.

King Tut's tomb was unearthed while digging for artifacts.

Looking through his binoculars, the bird flew away.

Developed by scientists at Wright State University, Dr. Moore claimed that the serum was an effective drug.

Another apparently masochistic student wrote, "After 10 minutes in boiling water, I transferred the flask to an ice-bath." Although she provided no details about her injuries, I hope that her grade reflected her dedication to biology. Finally, a particularly creative student wrote this sentence in a report: "After closing the incision, the chimp relaxed." I wonder where that ape went to medical school.

Keep Related Words Together

*Of all the faults found in writing, the wrong placing
of words is one of the most common, and perhaps it
leads to the greatest number of misconceptions.*
    —William Cobbett

The more quickly you get from A to B, the more likely your readers are to see the relationship of A and B. This requires that you keep related words together. If you fail at this, you'll probably confuse the reader. For example, consider this sentence that was included in a laboratory report:

Here are some suggestions for handling the buffer problems from Sigma Chemical Company.

This sentence confuses readers: Did Sigma provide suggestions for handling the problems, or did Sigma cause the problems? To eliminate this confusion, rewrite the sentence to say what you mean:

Sigma Chemical Company sent us some suggestions for solving the buffer problems.

Poorly organized sentences are often funny. For example, a recruiting brochure for a prominent university claimed that its biology department has "a botanical garden, electron microscope, and on-line computer connections in every teaching lab." A botanical garden in every lab? Similarly, a student wrote in her journal, "In the laboratory is a report written by Bill and an aquarium." Quite a talented aquarium.

Word order is also important. Consider the different meanings that result from reversing two words in the following sentences:

We almost lost our entire sample.

We lost almost our entire sample.

Both of these sentences are grammatically correct, but only one accurately describes the situation. Similarly, the order of words can create confusion or absurdity:

> We tested the patients using this procedure.

> Randy asked his students while in lab to watch the video.

> The marine biologist watched the sea gull wearing a bathing suit.

> Watson and Crick completed their paper while traveling on the back of a notebook.

> The table was used by the scientist with a broken leg.

> Gould will discuss how evolution has influenced the use of fossils in the classroom at noon.

> The patient left the hospital in good condition.

If you occasionally write sentences like these, don't feel too bad—so does everyone. For example, this caption appeared beneath a picture on the sports page of the *Washington Star-News* on June 12, 1976:

> Pele, the star of Team America, soothes the ankle he sprained in a scrimmage with an ice pack.

Many readers probably wondered why Pele was scrimmaging with an ice pack rather than with other soccer players. A more remarkable caption appeared in the premier issue of *Baylor: The News Journal of Baylor University* in January 1988:

> Former Baylor trustee Mrs. Dorothy Kronzer of Houston greets Jehan Sadat, widow of Egyptian President Anwar Sadat, who was the featured speaker at the fall Baylor University Forum for Distinguished Lecturers.

This was quite a debut for the *News Journal*. Readers must have wondered what strings Baylor University had to pull to get Anwar Sadat as a featured speaker in 1988, 7 years after his death. An equally remarkable story appeared on the front page of the October 3, 1985 issue of the *Dallas Morning News*:

> For the ninth time in less than two months, a young Indian man on the Wind River Reservation in Central Wyoming has committed suicide, officials said Tuesday.

Similarly, a local cafeteria posted a sign warning patrons, "Shoes and a shirt are required to eat in this cafeteria." Apparently my socks can eat anywhere they choose.

Be Careful with Pronouns

You'll confuse readers if you use indefinite pronouns such as "it" and "them" that lack a clear antecedent. For example, the sentence, *"Planting marigolds near cucumbers will keep the rabbits away from them"* is confusing: Will rabbits stay away from the marigolds, the cucumbers, or both? Here are other examples:

Randy asked Darrell if he could dissect the leaf.

When we tried to follow the instructions on the package, we burned it.

Not placing modifiers near words they modify can produce funny—even embarrassing—sentences. Comedians use this trick often. For example, recall Groucho Marx's famous lines from *Animal Crackers*: "One morning I shot an elephant in my pajamas. How he got in my pajamas I don't know." Biologists unknowingly become comedians when they ignore this principle of good writing. For example, few people rushed to respond to this advertisement:

Free information about VD—to get it call 873-2655.

Avoid Clichés

As soon as certain topics are raised, . . . no one seems able to think of turns of speech that are not hackneyed: prose consists less and less of words chosen for the sake of their meaning, and more and more of phrases tacked together like the sections of a prefabricated hen-house.
—George Orwell

A cliché is an expression so overworked that it has become an automatic way of getting around the main business of writing, which is to suit each word to the meaning at hand. The habitual user of cliché brands himself as a lazy writer.
—Frederick Crews

Modern writing at its worst . . . consists in gumming together long strips of words which have already been set in order by someone else and making the results presentable by sheer humbug.
—George Orwell

Clichés such as *quantum leap, foregone conclusion, marked contrast, wave of the future, acid test, better late than never, few and far between, first and foremost, it goes without saying, last but not least, bottom line, tried and true, state-of-the-art, advanced technology, too numerous to mention,* and *by the same token* are trite, overused expressions that cover absence of thought and laziness, and make for dull reading and listening.[7]

Use your own words, not the clichés of others. Avoid clichés ~~like the plague.~~

Write Positively

The cliché you probably remember about this "rule" is that you should not use two negative words in one sentence—that is, "don't use no double negatives." Although double negatives were used often by Chaucer, Shakespeare, and many other great writers of the past, double negatives are now considered unacceptable. Therefore, avoid expressions such as *haven't scarcely, can't help but,* and *not hardly any.*

Our minds function best when presented with information in a positive form. Therefore, say what is, not what isn't. Describe what happened or what you observed, not what you didn't observe or don't know. For example, instead of writing that the sample was not large enough, write that it was too small. Similarly, write that the rats were always sick, not that they were never healthy. Finally, instead of telling someone, "Pressure must not be lowered until the temperature is not less than 80°C," write, "Keep the pressure constant until the temperature is at least 80°C." Other examples of negative writing common in biology papers include the following phrases (negatives are in italics; preferred forms are in roman type):

not many	few	*not the same*	different
not different	alike	*did not*	failed to
did not remember	forgot	*does not have*	lacks
not able	unable; cannot	*not certain*	uncertain
did not accept	rejected	*did not allow*	prevented
did not consider	ignored	*did not stay*	left
not known	unknown	*not on time*	late
not possible	impossible	*not sure*	unsure
not aware	unaware		

Finding and changing negative phrases is easy. Use the "find" feature of your word-processing program to find every use of "not" in your paper; then ask yourself what word could replace the negative phrase. Such revisions are impor-

---

[7]As I was writing this section I happened to turn the television on to a sports show. The "student athlete" being interviewed rattled off about 10 *uhs* and 15 *you knows* that separated clichés such as, *"We've gotta give it 110%," "We don't know the meaning of the word quit," "I gave it all I got," "They came ready to play,"* and *"We'll just have to go out there and see what happens."* Curiously, he was wearing an "Albert Einstein" t-shirt.

tant because positive writing persuades people, while negative writing nags at and turns people off.

Double negatives usually leave writers dizzy from having to do mental cartwheels. Therefore, ~~don't write in the negative~~ write positively.

Avoid Abbreviations and Foreign Words

Abbreviations often confuse readers if they're not defined. For example, "MS" can mean mitral stenosis to a cardiologist, multiple sclerosis to a neurologist or virologist, or manuscript to an editor. If you must use abbreviations, be sure to define them.

Most English-speaking people understand English more readily than they do other languages such as Latin or Greek. For example, we know what *midbrain* means but are probably not too sure about *mesencephalon*. Therefore, write only in English so that you avoid the pedantic mumbo-jumbo created by foreign words like these:

a fortiori	with stronger reason
a priori	from cause to effect
ab initio	from the beginning
ad hoc	for this particular purpose
ad libitum	at will
infra dignitatem	undignified
loco citato	in the place cited
de facto	in fact, actually
in toto	entirely, completely
non sequitur	it does not follow
per se	of itself
per diem	daily
sine qua non	necessity

Unless instructed to do otherwise, also avoid Latin abbreviations such as:

cf.	confer	compare
et al.	et alii	and other people; use this abbreviation only when instructed to do so by a journal to shorten the list of authors of a reference. For example," . . . according to methods described by Burton *et al.* (1991)."

etc.	et cetera	and the rest, and so forth; using *etc.* means only that the list is incomplete. For example, consider the sentence, "For lab this week, bring a scalpel, ruler, *etc.*" What should you bring? A chain-saw? A goat? Use *etc.* only to escape having to repeat all the items of a list already given. Avoid *etc.* at all other times; it suggests that you either don't know what you're writing about or that you know but can't be bothered to tell the reader.
e.g.	exempli gratia	for example
et seq.	et sequentes	and the following
ibid.	ibidem	in the same place
i.e.	id est	that is; just say "that is" or "meaning that." Most people don't know what "i.e." means anyway.
infra dig.	infra dignitatem	undignified
loc. cit.	loco citato	in the place cited
N.B.	nota bene	note well
q.v.	quod vide	which see
viz.	videlicet	namely

Define all abbreviations that you use. For example, write "namely" instead of *viz*, and write "about" instead of *circa* or *ca*. If you insist on using foreign words or Latin abbreviations, know their meaning and do not mix them with English in the same phrase.

Express Similar Ideas in Similar Ways

Actors and actresses often bicker about who will get top billing in an upcoming movie—whose name will appear in ten-inch letters and whose will appear in half-inch letters. Something analogous occurs in writing when two or more similar ideas fight for our attention. To avoid confusing their readers, writers must express parallel ideas in parallel ways—either with all verbs, all prepositional phrases, or all adjectives. This is called parallelism. For example, consider how this famous quotation loses its punch when its ideas aren't treated equally:

   . . . that all men . . . have certain inalienable rights: . . . life, liberty, and to pursue happiness.

Despite the legendary struggles that Herbert Hoover had with writing as a student at Stanford, we see here an effective use of parallel structure to show the parallel relationships between coordinate ideas:

The great liability of the engineer compared to men of other professions is that his works are out in the open where all can see them. His acts, step by step, are in hard substance. He cannot bury his mistakes in the grave like the doctors. He cannot argue them into thin air or blame the judges like the lawyers. He cannot, like the architects, cover his failures with trees and vines. He cannot, like the politicians, screen his shortcomings by blaming his opponents and hope that the people will forget. The engineer simply cannot deny that he did it. If his works do not work, he is damned.[8]

This paragraph included in "More Is Not Merrier" by Andy Rooney also shows an effective use of parallelism:

We've already ruined all the rivers from the Yangtze to the Mississippi. Do you know of a lake you can drink from? Lake water was all drinkable before we started dumping our garbage, our sewage, and our commercial waste in it. Now we've starting ruining our oceans and, big as they are, they'll be seas of slop before long.

Many writers have much trouble with parallelism. Here are examples of those troubles from papers written by biology students:

I admire dolphins for their intelligence, energy, and because they are beautiful.

The description of the field site was both accurate and it was easy to read.

The limitations of this method are its limited sensitivity and that it is dependent on high humidity.

The typical biology student is motivated, a blond, and has blue eyes.

Revising these sentences improves their readability:

I admire dolphins for their intelligence, energy, and beauty.

The description of the field site was both accurate and readable.

The limitations of this method are its limited sensitivity and its dependence on high humidity.

The typical biology student is motivated, blond, and blue-eyed.

Parallel descriptions of ideas are usually linked with *and* or *or*. Locate and correct them with the search command of your word-processing program. To

[8]Hoover, Herbert, 1951. *Memoirs of Herbert Hoover, Vol. 1: Years of Adventure.* New York: Macmillan, pp. 132–133.

remember how to write them correctly, recall Lincoln's famous words at Gettysburg: ". . . government of the people, <u>by</u> the people, <u>for</u> the people . . .⁹

Meet the Expectations of Your Readers

The length-based assignments typical of most English classes have caused most students to adopt a "what do I say next?" approach to writing. This approach usually fails because the length of a paper is unrelated to the *quality* of a paper. To write a paper that helps you learn and communicate effectively, you must use a strategy based not on "what do I say next?" but rather on meeting readers' expectations. This strategy requires only that the *substance* of your writing (especially its context, action, and emphasis) appear in certain well-defined places within the *structure* of your writing. Although an understanding of readers' expectations is seldom included in writing courses and writing books, it is critical to effective communication. Once you know where your readers look for your message, you'll understand the many options you have to improve the chances of your reader finding your message. Understanding—and therefore being able to predict—how your writing will most likely be interpreted by readers will help move you away from a strategy of "not producing enough words" to a strategy of grasping, developing, and communicating ideas. It will help you write better every time you write, whether it be in a biology class or at any of life's many other rhetorical tasks. Moreover, since writing is inextricably linked with thinking, improving either one of these will necessarily improve the other.

If you want readers to grasp what you mean, then you must understand what they need as they read and interpret your paper. This involves understanding what parts of sentences and paragraphs readers look to for emphasis. In speaking, there are many ways to indicate stress. For example, we can alter the volume, speed, or accent of our speech to help readers understand our message. In writing, most emphasis results from the structure of our writing. That structure determines how your paper hangs together, while the substance contains the information that you want to communicate. To show this, consider these two ways of presenting information about the growth of roots:

t (time) = 0.0 days, h (height) = 3.1 mm; t = 6 hours, h = 5.8 mm; t = 12 hours, h = 11 mm; t = 24 hours, h = 20 mm; t = 48 hours, h = 43 mm; t = 96 hours, h = 79 mm

⁹The sentence from which these words are taken shows that a long sentence can be powerful. The sentence is 82 words long and is one of the most powerful sentences in the English language: "*It is rather for us to be here dedicated to the great task remaining before us—that from these honored dead we take increased devotion to that cause for which they gave the last full measure of devotion—that we here highly resolve that these dead shall not have died in vain—that this nation, under God, shall have a new birth of freedom—and that government of the people, by the people, for the people, shall not perish from the earth.*" Interestingly, the Gettysburg Address, one of the most famous speeches in American history, took only two minutes for Lincoln to recite.

time (hours)	height (mm)
0	3.1
6	5.8
12	11
24	20
48	43
96	79

Although each format presents the same information, the second format is easier to interpret because it gives you an easily perceived context (time) in which you can interpret the important information (height). The context appears on the left in a regular pattern, while the results on the right are in a less obvious pattern, the discovery of which is the point of the chart. You prefer this arrangement because you read from left to right: You prefer to get the familiar information first, followed by the new, important information. Any other arrangement slows your understanding because it puts information where you do not expect it.

Perceiving and absorbing information that we read requires energy. When we read, we have only a limited amount of this energy that we use in a zero-sum game: Any energy used to understand the paper's structure is unavailable to understand the paper's substance. If readers use an excess amount of energy to discern the structure of what you write, such as the energy required to understand the meaning of big words or how one sentence relates to a previous sentence, they have little or no energy left to find your message. This defines bad writing: By forcing readers to wade through confusing words and constructions, you unravel the structure of your writing and hide your meaning. This doesn't mean that your writing is impossible to understand; it only means that the reader is less likely to understand what you're trying to say. However, if you put words and connections where readers expect them to appear, the reader can devote more of his or her energy to understanding your ideas. These changes produce subtle and remarkably significant effects, not only in your thinking, but also in your writing.

Follow the Subject as Soon as Possible with Its Verb

Just as scientists expect certain kinds of information in particular sections of a scientific paper, readers have fixed expectations about where they will find important information in a sentence or paragraph. Consequently, they search for that information in those places. By knowing where these places are, you (rather than the person reading or grading your paper) can identify and correct vagueness in your writing by perceiving problems with the structure of your writing. Consider this example:

The smallest of the URF's (URFA6L), a 207-nucleotide (nt) reading frame overlapping out of phase the NH_2-terminal portion of the adenosinetriphosphatase (ATPase) subunit 6 gene has been identified as the animal equivalent of the recently discovered yeast H^+-ATPase subunit 8 gene. The functional significance of the other URF's has been, on the contrary, elusive. Recently, however, immunoprecipitation experiments with antibodies to purified, rotenone-sensitive NADH-ubiquinone oxidoreductase [hereafter referred to as respiratory chain NADH dehydrogenase or complex I] from bovine heart, as well as enzyme fractionation studies, have indicated that six human URF's (that is, URF1, URF2, URF3, URF4, URF4L, and URF5, hereafter referred to as ND1, ND2, ND3, ND4, ND4L, and ND5) encode subunits of complex I. This is a large complex that also contains many subunits synthesized in the cytoplasm. . . .

Why is this paragraph hard to read? Most people would say that it requires background knowledge, while others would mention its technical vocabulary. These issues are only a small part of the problem. Here's the passage with its difficult words removed:

The smallest of the URF's, an [A], has been identified as a [B] subunit 8 gene. The functional significance of the other URF's has been, on the contrary, elusive. Recently, however, [C] experiments, as well as [D] studies, have indicated that six human URF's [1–6] encode subunits of Complex I. This is a large complex that also contains many subunits synthesized in the cytoplasm. . . .[10]

This version is more readable but still difficult to understand. Although knowing the meanings of "URF" ("uninterrupted reading frame," a segment of DNA that could encode a protein, although no such protein has been identified) and ATPase and NADH oxidoreductase (enzymatic complexes involved in cellular energetics) provides some sense of comfort, it doesn't remedy the confusion. The reader is hindered by more than just the jargon and a lack of background information. What, then, is the problem?

Look again at the first sentence of the original passage. Although it is long (42 words), that's not the real problem; long sentences need not be hard to read. The first sentence of the passage is difficult to read not because it's a long sentence, but because it presents information in places where readers don't expect the information. For example, the sentence's subject ("the smallest") is separated from its verb ("has been identified") by 23 words, more than half the sentence. This violates a fundamental expectation of readers, namely that the subject should be followed immediately by the verb. Anything significant placed between the subject and verb interrupts the reader, causing him or her to lessen

---

[10]From Gopen, George D., and Judith A. Swan. 1990. "The Science of Scientific Writing." *American Scientist* 78: 550–558. An excellent article about writing to meet readers' expectations.

the sentence's importance. Here's how the writer could have improved the sentence:

> The smallest of the URF's is URFA6LK, a 207-nucleotide (nt) reading frame overlapping out of phase the NH2-terminal portion of the adenosinetrisphosphatase (ATPase) subunit 6 gene; it has also been identified as the animal equivalent of the recently discovered yeast H+-ATPase subunit 8 gene.

Similarly, if the intervening material is trivial to the idea, then rewrite the sentence like this:

> The smallest of the URF's (URFA6L) has been identified as the animal equivalent of the recently discovered yeast H+-ATPase subunit 8 gene.

This sentence gets directly to its point. However, only the author could tell us which of these versions more accurately reflects his or her intentions.

Build Your Writing around Specific Nouns and Strong Verbs

[M]ake your words forceful, compact, and
energetic. . . . A few strong, carefully selected
words will deliver your message with the kick of a
mule.
—Mark S. Bacon, *Write Like the Pros*

Suit the action to the word, the word to the action.
—William Shakespeare

As to the adjective: when in doubt, strike it out.
—Mark Twain

Strong verbs are the engines that move sentences, no matter if you're talking about a trashy novel, the works of Hemingway, or a scientific masterpiece. All good writers energize their writing by anchoring sentences with strong verbs and by deleting smothering adjectives. Many of these adjectives diminish the meaning of precise words. Others, such as *rather* (*rather* important) and *almost* (*almost* unique) can be contradictory and should therefore be omitted. Finally, other adjectives intensify the meaning of words that are already emphatic. For example, instead of writing that something is *very important*, write that it is *critical* or *crucial*. Or just delete *very* and say that it is *important*. Choose specific nouns that need few adjectives.

Few things improve a sentence more than a well-chosen verb. You'll energize your writing by (1) using verbs that express action, (2) not making nouns out of strong, working verbs, and (3) preferring active voice.

Use verbs to express the action of every clause or sentence. Rambling, unwieldy sentences often hang on wimpish, inert verbs such as *exist, occur,* and forms of *to be* and *to have.* These verbs bring the motion of a sentence to a standstill. Replacing inert verbs with action verbs always strengthens your writing. To show this, consider this question:

What would be the students' reception accorded the introduction of such an experiment?

In earlier chapters you learned how to revise this question by removing unnecessary words. Consider this revision:

How would the students receive such an experiment?

The revision is easier to read than is the original question. Why? The revision is shorter, but its reduced length is a result rather than a cause of the improvement. To understand why the revision is easier to read, study the original question and determine what action is occurring. Action words in the first question are *be, reception, accorded,* and *introduction,* while in the second question *receive* is the only action in the sentence. This is a critical distinction because readers expect to find action in the verb. In the original question, *accorded* sounds like action but makes no sense as action. This forces readers to divert energy from understanding your message to deciphering its structure. Consequently, when the complexity increases moderately, the chance of misinterpretation or noninterpretation increases dramatically.

Inert verbs that imply rather than state action are common in scientific writing. For example, consider these sentences:

INERT: Ruska performed the development of the electron microscope.

IMPROVED: Ruska <u>developed</u> the electron microscope.

INERT: Watson and Crick made the decision to publish their discovery in *Nature.*

IMPROVED: Watson and Crick <u>decided</u> to publish their discovery in *Nature.*

In these examples, *performed* and *made* imply action and therefore require other verbs to complete the idea. For example, *performed the development of* means *developed,* and *made the decision* means *decided.*

Improve your sentences by asking yourself these questions:

What's happening here? What did I intend the action to be? Does a verb announce that action?

Use a verb to announce the action of the sentence. By doing so, you help ensure that readers understand your message. This is another example of how rewriting is rethinking and how revision is invention. Indeed, weak verbs allow writers to mechanize their ideas as narration or recitation of facts rather than to explain their thoughts. Force yourself to consider what is happening by stating the action in a strong verb that explains your thoughts.

Don't make nouns out of strong, working verbs. Many scientists make nouns out of strong verbs. This is especially common among writers who are uncomfortable with simple, direct statements. To show this, consider the following sentence:

Authorization for the experiment was given by the professor.

This writer has turned *authorize*, a strong verb, into *authorization*, a weak noun that slows the pace of the sentence. Similar *-ion* nouns such as *determination, consideration*, and *conclusion* produce unnecessarily long and dull sentences —long because their weak verbs attract other unnecessary words and dull because their verbs are abstract. Look at how much more convincing and direct the sentence becomes when you rewrite the sentence around the action verb:

The professor authorized the experiment.

Here are some other examples of sentences that improve when their smothered verb is changed to an action verb:

We conducted a study of the cells.
We **studied** the cells.

A need exists for greater proposal selection efficiency.
We must **select** proposals more efficiently.

Martha has expectations of being accepted to medical school.
Martha **expects** to get into medical school.

Our discussion concerned the evolution of the dorsal fin.
We **discussed** the evolution of the dorsal fin.

Here are some other examples of smothered verbs and their strong counterparts:

Smothered	*Strong or Improved*
draw conclusions	conclude
make assumptions	assume
take action	act
make use of	use
bring to a conclusion	conclude
make a determination	determine
have an influence on	influence
give consideration to	consider

Smothered	*Strong or Improved*
arrive at a decision	decide
achieve purification	purify
are found to be in agreement	agree
bring to closure	end/finish
is indicative of	indicates
make mention of	mention
institute an improvement in	improve
present a summary	summarize

Substituting strong verbs for smothering verbs shortens and clarifies sentences by pushing ideas to readers. This explains why strong verbs are a hallmark of effective writing.

Prefer active voice. In writing, voice refers to the relationship of a verb to its subject. If the subject of a verb does the action, the sentence is written in *active voice*. If the subject receives the action, the sentence is written in *passive voice*.

Many scientists replace active, personalized storytelling with a passive, abstract style of writing based on *to be* verbs such as *is, was*, and *were*. In these sentences, the subject is acted upon. Here are some examples of passive voice:

It was suggested that the experiment be terminated. (Who suggested this? Terminated by whom?)

It was determined that information was insufficient for the biologist to recommend specific action on the question of nutritional needs of the elderly in the designated study.

A cracker is wanted by Polly.

Note that the lead verb in each of these sentences is *was* or *is* and identifies no one responsible for the action. Passive voice typically produces long sentences that sap readers' strength and often sound pompous. Furthermore, passive voice is usually weak and unconvincing, suggesting that scientists were acted upon rather than that scientists acted. With passive voice, it seems as if the writers are reporting revelation, not information, because passive voice implies that a force —the amorphous "it" —guided their work.

Active voice is a writing style in which the subject acts. Changing the sentences listed above to active voice produces the following sentences:

I suggested that we end the experiment.

The biologist lacked enough information to recommend nutritional improvements for the elderly included in the study.

Polly wants a cracker.

Scientific Writing, Objectivity, and Passive Voice

First-person pronouns such as "I" and "we" disappeared from scientific writing in the United States in the 1920s, when today's inflexible and impersonal style of scientific writing began to dominate science and technology. Since then, scientists have used the anonymity of passive voice to make themselves appear as modest, passive, and objective observers. This is unfortunate because it greatly diminishes communication. For example, when speaking, most scientists provide information in the way we ordinarily expect to receive it—as a narrative:

> We wanted to understand how penicillin affects growth of bacteria. To do this, we grew bacteria in the presence of varying concentrations of penicillin. We learned that penicillin inhibits growth of bacteria.

Such a narrative isn't something any of us have to think about; we speak like this all the time. And when we aren't telling these kinds of stories, we're listening to them. Narrative communication is easy for everyone. However, compare this with the abstraction of passive voice:

> The growth of bacteria was studied. Bacteria were grown in the presence of varying concentrations of penicillin. It was learned that bacterial growth is inhibited by penicillin.

Nobody in the real world communicates like this. Consequently, the insistence of many scientists to write only in passive voice forces readers to shift into a foreign mode of communication. Passive voice is hard to write and even harder to read because, especially in reports of experiments, it doesn't reflect what really happened. More importantly, the abstraction of passive voice often provides less information than a narrative using active voice because it removes the feelings, the flavors, the juice, and sometimes even the substance of what we did.

The notion that passive voice ensures objectivity is artificial because objectivity has nothing to do with one's writing style or use of personal pronouns. Objectivity in science results from the choice of subjects, facts that you choose to include or omit, sampling techniques, and how you state your conclusions. Scientific objectivity is a personal trait unrelated to writing. Relying on third-person (for example, "the author") achieves no modesty, and discarding *I* and *we* merely leads to awkward, weak, and indirect writing.

Passive voice can be useful to writers. For example, passive voice is effective when you want to stress "what was done" rather than "who did it," as in the sentence "Darwin's *The Origin of Species* was published in 1859." Passive voice is also useful when you want to avoid accountability. For example, embarrassed politicians report that "funds were found to be missing," not that "I stole the money." Rather than fool anyone, the excessive use of passive voice usually just bores readers. Passive voice is useful for adding variety, softening commands, avoiding responsibility, avoiding embarrassment, and slowing the pace of writing and reading. Unless you have one of these reasons for using passive voice, choose active voice.

Some may be shocked by the bluntness of these sentences. Don't worry about people with this concern. Active voice almost always improves writing because it shortens sentences, improves readability, makes analyses more incisive, increases understanding, and more effectively reveals the depth of a scientist's logic. Active voice also produces direct, clear sentences that involve readers. Conversely, passive voice is an impersonal, and often boring, style of writing.

Most scientists use passive voice either out of habit or to make themselves seem scholarly, objective, or sophisticated. The result, however, is boring writing. For example, consider these sentences:

There was considerable erosion of the land by the floods.

The patients were examined by the physicians.

Papers are written by students.

All of these sentences are grammatically correct. However, all are based on weak verbs such as *is, are*, and *were*—derivatives of *to be*. Notice how these sentences improve when you change passive voice to active voice:

Floods eroded the land.

The physicians examined the patients.

Students write papers.

These sentences are shorter, more direct, and communicate better than those written in passive voice.

Contrary to the implication of passive voice, science is a personal activity done by people, not machines. Therefore, report your work as you did it. Look at these examples of active voice taken from consecutive sentences of an article in *Science*, a prestigious journal:

. . . we want . . . Survival gives . . . We examine . . . We compare . . . We have used . . . We merely take . . . They are subject . . . We use . . . Effron and Morris (3) describe . . . We observed . . . We might find . . . We know . . .

Similarly, look at how this paragraph from a paper published in *The New England Journal of Medicine* uses active voice to communicate directly with readers (boldface added):

Our findings concerning the overall incidence of acute and chronic GVHD are not appreciably different from those of other groups. Patients with Grades 0 through II acute GVHD fared better than those with more severe involvement, the latter group having a higher incidence of fatal infection. In this series, **we** found no association between acute GVHD and a reduced risk of posttransplantational relapse. However, as in other recent reports, we did not find an association between the presence of chronic GHVD and a low relapse rate.[11]

Finally, note how James Watson and Francis Crick used active voice and a simple, personal, get-to-the point writing style to open their monumental paper describing the structure of DNA:

> We wish to suggest a structure for the salt of deoxyribose nucleic acid (D.N.A.).

The notion that "I" and "we" somehow make science undignified is foolish and hobbles science. After all, all research is done by someone, and every paper, book, and essay has someone's name listed as the author. Most editors prefer that writers use "I" or "we" to describe research. For example, the *American National Standards for the Preparation of Scientific Papers for Written or Oral Presentations*, which includes an impressive list of organizations represented by its views, states: "When a verb concerns action by the author, the first person should be used, especially in matters of experimental design."[12] If the purpose of your paper is to tell what you did and observed, then you should appear at the beginning of sentences and clauses. Insistence on using "the authors" or "the writers of this paper" is pompous, distant, and stuffy.

Reject the argument that scientists should use only passive voice. Indeed, great scientists such as Einstein, Faraday, Watson, Crick, Curie, Darwin, Lyell, Freud, and Feynman communicated their brilliant ideas by preferring active over passive voice.

Science is the great adventure of our time. Don't suffocate it with passive, abstract writing.

Place at the End of a Sentence the Material You Want the Reader to Emphasize

So far in this chapter you've learned the importance of carefully chosen words and phrases for communication. These words and phrases are the ingredients of the next stage of effective writing: strong, concise sentences. Being able to write these sentences depends largely on knowing how to start and end a sentence.

Readers naturally emphasize the material at the end of a sentence. Consequently, they expect the most important idea—the idea that you want to emphasize—in the "stress position" at the end of a sentence. Periods end stress

[11]Brochstein, Joel A., Nancy A. Kernan, Susan Groshen, Constance Cirrincione, Brenda Shank, David Emanuel, Joseph Laver, and Richard J. O'Reilly. 1987. "Allogeneic Bone Marrow Transplantation after Hyperfractionated Total-Body Irradiation and Cyclophosphamide in Children with Acute Leukemia." *The New England Journal of Medicine* 317: 1618–24.

[12]American National Standards Institute (ANSI). 1979. Z39. 16-1979, *American National Standards for the Preparation of Scientific Papers for Written or Oral Presentations.* New York: American National Standards Institute, p. 12.

positions, while colons and semicolons indicate secondary stress positions in the sentence. If you fill the stress position with a word that might be emphasized, readers sense that you want to emphasize that word. Conversely, if you put emphatic material of a sentence anywhere but in the stress position, the reader may find the stress position occupied by material that you don't want to emphasize. Nevertheless, readers will probably emphasize the "imposter" material, killing your chances of influencing their interpretation. Alternately, if the stress position is filled with material that can't be logically emphasized, the reader must guess from all the other possibilities in the rest of the sentence what you want to emphasize. Moreover, the longer the sentence, the greater the chances of your reader choosing the wrong information for emphasis. Such guessing causes you to lose control of the reader's interpretation of your writing.

Put at the end of a sentence the newest or most significant information—the information that you want to stress and information that you will expand on in your next sentence. When you introduce an important term for the first time, design the sentence so that the term appears at the end of the sentence, even if you must invent a sentence just to define or emphasize that term. For example, look at how this paper in *The New England Journal of Medicine* introduces readers to lymphokine-activated killer cells:

> We have previously described a method for generating lymphocytes with antitumor reactivity. The incubation of peripheral-blood lymphocytes with a lymphokine, interleukin-2, generates lymphoid cells that can lyse fresh, noncultured, natural-killer-cell-resistant tumor cells but not normal cells. We have termed these cells lymphokine-activated killer (LAK) cells.

Similarly, look at how Charles Darwin emphasized evolution in the closing words of Chapter 21 of *The Descent of Man*:

> We must, however, acknowledge, as it seems to me, that man with all his noble qualities, still bears in his bodily form the indelible stamp of his lowly origin.

Another simple way to effectively end a sentence is to delete unnecessary words until the stressed information is at the end of the sentence. For example:

> The teaching assistant said that the ocular and objective lens determine magnification of the specimen during use of the microscope.

> The teaching assistant said that the ocular and objective lens determine magnification of the specimen.

Knowing to place important information in the stress position of a sentence will help you write concise, emphatic sentences. It will also help you identify sentences whose length detracts from your message. Longer sentences have more possible stress positions, thus explaining why longer sentences often communicate less effectively than shorter sentences. A sentence is too long when there is more than one candidate for a stress position.

Place Familiar Information at the Beginning of Sentences

For me, the big chore is always the same—how to
begin a sentence, how to continue it, how to
complete it.
    —Claude Simon, on winning the Nobel Prize in
        1985

Just as readers naturally emphasize the material at the end of a sentence, so too do they search for familiar information at the beginning of a sentence. Readers use the beginning of a sentence to link them with previous information and to provide context for upcoming material. Therefore, use the beginning of a sentence to put readers in familiar territory; this helps readers see the logic of your argument. If you repeatedly begin sentences with new information and end them with old information, you force readers to carry the new information further into the sentence before it can be put into context. This consumes much of the reader's energy, leaving them with little energy for understanding your message.

How you begin sentences is critical to how readers will understand them. The secret to a clear and readable sentence lies in the first five or six words of the sentence. Remember this: If you have to go more than six or seven words into a sentence to get past the subject and verb, or if subject is not one of those words, revise the sentence so that the main ideas appear as subjects and the actions appear as verbs. If you consistently organize the subject around a few concepts and then express that subject's action with a strong, precise verb, your readers will understand what you are writing about. If you do not do this, your writing will seem not just unfocused but weak and anticlimactic.

Each sentence should teach the reader something new. To ensure this, design your sentences as good teachers design their lectures—by using simple, straightforward language to expand a previous idea or to connect new ideas with what the reader already knows. Provide context before asking readers to consider anything new. Do this by putting the reader in familiar territory: Start sentences with ideas that you have already mentioned, referred to, or implied. For example, notice how this writer uses the phrase *this energy* to show the relationship between the energy in a slice of apple pie and the potential uses of that energy:

A slice of apple pie contains about 1.5×10^6 J (365 Cal) of energy. **This energy** is enough energy for a woman to run for almost an hour or for a typist to enter about 15,000,000 characters on a manual typewriter.

Note how Rachel Carson linked the sentences in this paragraph from *Silent Spring*, a masterpiece of biological literature:

From the green depths of the offshore Atlantic many paths lead back to the coast. They are paths followed by fish; although unseen and intangible, they are linked with the outflow of waters from the coastal rivers. For thousands upon thousands of years the salmon have known and followed these threads of fresh water that lead them back to the rivers, each returning to the tributary in which it spent the first months or years of life. So, in the summer and fall of 1953, the salmon of the river called Miramichi on the coast of New Brunswick moved in from their feeding grounds in the far Atlantic and ascended their native river. In the upper reaches of the Miramichi, in the streams that gather together a network of shadowed brooks, the salmon deposited their eggs that autumn in beds of gravel over which the stream water flowed swift and cold. Such places, the watersheds of the great coniferous forests of spruce and balsam, of hemlock and pine, provide the kind of spawning grounds that salmon must have in order to survive.[13]

Don't confuse restatement with explanation. Avoid starting sentences with phrases such as, "*That is,*" and delete sentences preceding those beginning with, "*In other words.*" Similarly, don't start sentences with, "It is interesting to note that . . ."—you'll only tempt the reader to find it dull. Rather, select words to *make* it interesting. Finally, don't say, "I might add that . . ."—just add it.

Use the structure of a sentence to persuade your readers of the relative values of the sentence's contents. Put at the beginning of a sentence the familiar information that links backward. Similarly, put new information that you want the reader to emphasize at the end of the sentence where the reader naturally exerts the greatest emphasis when reading the sentence. Beginning with the exciting material and ending with what readers already know leaves readers disappointed. Save the best for last—don't start with the ice cream and end with the spinach.

A Final Word About Choosing Your Words

Whatever one wishes to say, there is one noun only by which to express it, one verb only to give it life, one adjective only which will describe it. One must search until one has found them, this noun, this verb, this adjective, and never rest content with approximations, never resort to trickery, however happy, or to illiteracies, so as to dodge the difficulty.
    —Guy de Maupassant, *Pierre et Jean*

[13]Carson, Rachel. 1962. *Silent Spring.* Boston: Houghton.

What Do I Need to Know about Grammar?

Many people are uneasy about anything associated with grammar. Others such as Panini (a 5th century B.C. Indian scholar who wrote one of the world's first grammar books) have claimed that "who knows my grammar knows God." Although most writers wouldn't go quite that far, it's clear that grammar means different things to different people. Why all the fuss? What does a biologist need to know about grammar to write well? And what about all of those rules?

Good grammar isn't as easy as rules invented centuries ago, repeated by editors unwilling to determine whether those rules comport with reality, taught by teachers who teach only what textbooks tell them, and ignored by the best writers everywhere. Thoughtlessly following all of the rules all of the time will offend no one but will deprive your writing of any flexibility. Indeed, some of the most effective writers are the best violators of rules. These writers *usually* follow grammatical rules—that's why their occasional violation of a rule is so noticeable. Effective writers use such violations for effect in their writing.

Some rules only cause problems if they're followed. For example, forget the adage, "vary your word choice." Don't try to substitute similar words for the exact word when the exact word is critical to the meaning of a sentence. If you use two words for the same concept, you risk having readers think that you mean two concepts. This is why papers written by professional educators often are so confusing. Rather than repeat the word *explanation*, educators often substitute words such as *symbolic modeling, precept, language symbols, words, narrative modeling*, and *instructions*. Instead of providing "elegant variation," this refusal to write precisely merely confuses readers.

Several other so-called "rules" lack substance and are ignored by all good writers. For example, ignore the rule about never beginning a sentence with *because, and*, or *but*. Also ignore the notion that you should never split an infinitive. This rule originated with eighteenth-century writers who reasoned that since you can't split the one-word Latin infinitives, you shouldn't split English infinitives. Ignore the "split infinitive rule" if it impedes communication —If you can communicate best by saying *to boldly go*, then say it.

Other rules are imperatives that we violate at the risk of seeming at least careless, at worst illiterate. These rules are observed by even the less-than-best writers. Don't break these rules:

Avoid double negatives.
Example: I have not got no data.

All words are pegs to hang ideas on.
   —Henry Ward Beecher

As you write, remember that words are the only tool you're given. Therefore, use them originally, directly, and carefully. Value their strength and diversity. And remember—someone is listening.

Avoid nonstandard verbs.
Example: I knowed that she was my lab partner.

Avoid double comparatives.
Example: Paramecium is more quicker than *Amoeba.*

Do not substitute adjectives for adverbs.
Example: The professor did work real good.

Do not use incorrect pronouns.
Example: Her and me made the media.

Don't let the subject disagree with the verb.
Example: They was in the lab.

What good, then, is grammar? Many scientists view grammar as a series of formal rules that are unnecessary as long as they can "get the message across." This attitude is typical of ineffective writers and may suffice as long as you have nothing important to write. However, when you have something important to say, an understanding of grammar is critical because in writing, language is the way words are used. To understand how to use words effectively, you must understand something about grammar because it functions as a series of signposts that, if followed, organizes your thoughts and improves your writing and learning. Grammar deals with how words function in sentences and therefore underlies what words mean. Writing is effective only when it communicates, and that occurs most efficiently when you understand a few principles of grammar. Grammatical conventions represent agreement, and agreement stimulates communication.

Words can do many things, and one must understand grammar to use words effectively. Weak grammar puts a greater burden on the reader to grasp your point. And the harder it is for others to get your point, the less likely you are to communicate well, no matter how good your point may be.

Understanding grammar helps you clearly think about and precisely express your ideas. It helps you say what you mean in an orderly and precise way, usually by eliminating common sources of confusion. Although grammar can't substitute for knowing what you want to say, it can save you time by eliminating hazards that block communication and thinking. However, you do not need to understand all of the nuances of grammar to use grammar effectively, just as you do not need to understand how a watch works to tell time.

John Tierney with Lynda Wright and Karen Springen

The Search for Adam and Eve[14]

Scientists have long searched for and speculated about our common ancestor. That search bewildered scientists and nonscientists alike for decades. However, in the late 1980s, molecular biologists at the University of California at Berkeley and Emory University in Atlanta came upon an answer — not by digging up bones in Africa but by untangling strands of DNA. Based on their data, these biologists published the date of our ancestor's birth. Although such a discovery was strikingly important, it remained a difficult subject to explain to the public. For example, how can scientists trace our ancestry to one person? And why is the tracer inherited only on the mother's side? To further complicate matters, geneticists and anthropologists disagreed about many of the issues comprising the story.

John Tierney, a freelance writer, accepted the challenge to write such an article for *Newsweek*. He and his colleagues began by researching the subject, after which Tierney attended the annual meeting of the American Anthropological Association to speak with scientists about the controversy. To reconcile the positions of geneticists and anthropologists, Tierney began at a point of agreement — the evolutionary place of chimpanzees — and then went on to describe a detective story of tracking human lineage to the Mitochondrial Mother. One editor at *Newsweek* asked Tierney to insert a billboard — a statement of

[14]From *Newsweek*, January 11, 1988 and © 1988, Newsweek, Inc. All rights reserved. Reprinted by permission.

the article's theme. That statement leads off the
second paragraph.

"The Search for Adam and Eve" appeared in
the January 11, 1988 issue of *Newsweek*. It won
the AAAS-Westinghouse Science-Writing Award.

Scientists are calling her Eve, but reluctantly. The name evokes too
many wrong images — the weakwilled figure in Genesis, the milk-
skinned beauty in Renaissance art, the voluptuary gardener in "Paradise
Lost" who was all "softness" and "meek surrender" and waist-length
"gold tresses." The scientists' Eve — subject of one of the most
provocative anthropological theories in a decade — was more likely a
dark-haired, black-skinned woman, roaming a hot savanna in search of
food. She was as muscular as Martina Navratilova, maybe stronger; she
might have torn animals apart with her hands, although she probably
preferred to use stone tools. She was not the only woman on earth, nor
necessarily the most attractive or maternal. She was simply the most
fruitful, if that is measured by success in propagating a certain set of
genes. Hers seem to be in all humans living today: 5 billion blood
relatives. She was, by one rough estimate, your 10,000th-great-grandmother.

When scientists announced their "discovery" of Eve last year, they
rekindled perhaps the oldest human debate: where did we come from?
They also, in some sense, confirmed a belief that existed long before the
Bible. Versions of the Adam-and-Eve story date back at least 5,000 years
and have been told in cultures from the Mediterranean to the South
Pacific to the Americas. The mythmakers spun their tales on the same
basic assumption as the scientists: that at some point we all share an
ancestor. The scientists don't claim to have found the *first* woman,
merely a common ancestor — possibly one from the time when modern
humans arose. What's startling about this Eve is that she lived 200,000
years ago. This date not only upsets fundamentalists (the Bible's Eve was
calculated to have lived 5,992 years ago), it challenges many evolutionists'
conviction that the human family tree began much earlier.

Eve has provoked a scientific controversy bitter even by the standards
of anthropologists, who have few rivals at scholarly sniping. Their feuds
normally begin when someone's grand theory of our lineage is
contradicted by the unearthing of a few stones or bones. This time,
however, the argument involves a new breed of anthropologists who
work in air-conditioned American laboratories instead of dessicated
African rift valleys. Trained in molecular biology, they looked at an
international assortment of genes and picked up a trail of DNA that led
them to a single woman from whom we are all descended. Most
evidence so far indicates that Eve lived in sub-Saharan Africa, although a
few researchers think her home might have been southern China.

Meanwhile, other geneticists are trying to trace our genes back to a scientifically derived Adam, a putative "great father" of us all. As is often the case, paternity is proving harder to establish: the molecular trail to Adam involves a different, more elusive sort of DNA.

The most controversial implication of the geneticists' work is that modern humans didn't slowly and inexorably evolve in different parts of the world, as many anthropologists believed. The evolution from archaic to modern *Homo sapiens* seems to have occurred in only one place, Eve's family. Then, sometime between 90,000 and 180,000 years ago, a group of her progeny left their homeland endowed apparently with some special advantage over every tribe of early humans they encountered. As they fanned out, Eve's descendants replaced the locals, eventually settling the entire world. Some "stones-and-bones" anthropologists accept this view of evolution, but others refuse to accept this interpretation of the genetic evidence. They think our common ancestor must have lived much farther in the past, at least a million years ago, because that was when humans first left Africa and began spreading out over the world, presumably evolving separately into the modern races. As the veteran excavator Richard Leakey declared in 1977: "There is no single center where modern man was born."

But now geneticists are inclined to believe otherwise, even if they can't agree where the center was. "If it's correct, and I'd put money on it, this idea is tremendously important," says Stephen Jay Gould, the Harvard paleontologist and essayist. "It makes us realize that all human beings, despite differences in external appearance, are really members of a single entity that's had a very recent origin in one place. There is a kind of biological brotherhood that's much more profound than we ever realized."

This brotherhood was not always obvious in Chicago two months ago, when the Eve hypothesis was debated by the American Anthropological Association. Geneticists flashed diagrams of DNA, paleoanthropologists showed slides of skulls and everyone argued with everyone else. "What bothers many of us paleontologists," said Fred Smith of the University of Tennessee, "is the perception that this new data from DNA is so precise and scientific and that we paleontologists are just a bunch of bumbling old fools. But if you listen to the geneticists, you realize they're as divided about their genetic data as we are about the bones. We may be bumbling fools, but we're not any more bumbling than they are." For all their quarrels, though, the two groups left Chicago convinced they're closer than ever to establishing the origin of modern humanity. To make sense of their bumbling toward Eden, it may be best to go back to one ancient relative accepted by all scientists. That would be the chimpanzee.

Until the molecular biologists came along, the role of the chimpanzee in evolution rested on the usual evidence: skeletons.

Scientists have relied on bones ever since the 1850s, when Darwin published his theory of evolution and some quarriers unearthed a strange skeleton in Germany's Neander Valley. Was the stooped apelike figure a remnant of an ancient race? Leading scientists thought not. One declared it a Mongolian soldier from the Napoleonic Wars. A prominent anatomist concluded it was a recent "pathological idiot."

But more skeletons kept turning up across Europe and Asia. Anthropologists realized that Neanderthal man was one of many brawny, beetle-browed humans who mysteriously disappeared about 34,000 years ago. These early Homo sapiens, incidentally, were not stooped (that first skeleton was hunched with arthritis). Nor did they fit the stereotype of the savage cave man. Their skulls were thicker than ours, but their brains were as large. Their fossils show that they cared for the infirm elderly and buried the dead. It seemed they might be our ancestors after all.

Meanwhile, fossil hunters in Asia more than a half century ago found the still older bones of Java man and Peking man, who had smaller brains and even more muscular bodies. These skeletons dated back as far as 800,000 years. Perhaps they represented evolutionary dead ends. Or perhaps they, too, were human ancestors, with their descendants evolving into modern Asians while the Neanderthals were becoming modern Europeans—a process of racial differentiation that lasted a million years. Either way, it appeared that all these ancient humans traced their lineage back to Africa, because that was the only place with evidence of humans living more than a million years ago. Stone tools were invented there about 2 million years ago by an ancestor named *Homo habilis* ("Handy Man"). Before him was Lucy, whose 3-million-year-old skeleton was unearthed in the Ethiopian desert in 1974. (Her discoverers celebrated by staying up all night drinking beer, and they named her after the Beatles' song that kept blaring on the camp's tape player, "Lucy in the Sky with Diamonds.") Lucy was three and a half feet tall and walked erect—not ape, not quite human. At some point her hominid ancestors had begun evolving away from the forebears of our closest relative, the chimpanzee.

But when? Most anthropologists thought it was at least 15 million years ago, because they had found bones from that era of an apelike creature who seemed to be ancestral to humans but not apes. Then, for the first time, geneticists intruded with contradictory evidence, led in 1967 by Vincent Sarich and Allan Wilson of the University of California, Berkeley. They drew blood from baboons, chimps and humans, then looked at the molecular structure of a blood protein that was thought to change at a slow, steady rate as a species evolved. There were major differences between the molecules of chimps and baboons, as expected, since the two species have been evolving separately for 30 million years. But the difference between humans and chimps was surprisingly small—so small, the geneticists concluded, that they must have parted company just 5 million years ago. Other geneticists used different techniques and came up with a figure of 7 million years.

Traditional anthropologists did not appreciate being told their estimates were off by 8 million or 10 million years. The geneticists' calculation was dismissed and ignored for more than a decade, much to Wilson's displeasure. "He was called a lunatic for 10 years. He's still sensitive," recalls Rebecca Cann, a former colleague at Berkeley who is now at the University of Hawaii. But eventually the geneticists were vindicated by the bones themselves. As more fossils turned up, anthropologists realized that the 15-million-year-old bones didn't belong to a human ancestor and that chimps and humans did indeed diverge much more recently.

Now Wilson, who won a MacArthur "genius grant" in 1986, is once again trying to speed up evolution. The Eve hypothesis, being advanced both by his laboratory and by a group at Emory University, is moving up the date when the races of humanity diverged—and once again Wilson faces resistance. Some anthropologists aren't happy to see Neanderthal and Peking man removed from our lineage, consigned to dead branches of the family tree. Wilson likes to remind the critics of the last fight. "They're being dragged slowly along," he says. "They'll eventually come around."

To find Eve, Cann first had to persuade 147 pregnant women to donate their babies' placentas to science. The placentas were the easiest way to get large samples of body tissue. Working with Wilson and a Berkeley biologist, Mark Stoneking, Cann selected women in America with ancestors from Africa, Europe, the Middle East and Asia. Her collaborators in New Guinea and Australia found Aboriginal women there. The babies were born, the placentas were gathered and frozen, and the tissue analysis began at Wilson's lab in Berkeley. The tissues were ground in a souped-up Waring blender, spun in a centrifuge, mixed with a cell-breaking detergent, dyed fluorescent and spun in a centrifuge again. The result was a clear liquid containing pure DNA.

This was not the DNA in the nucleus of the babies' cells—the genes that determine most physical traits. This DNA came from outside the nucleus, in a compartment of the cell called the mitochondrion, which produces nearly all the energy to keep the cell alive. Scientists didn't learn that the mitochondrion contained any genes until the 1960s. Then in the late 1970s they discovered that mitochondrial DNA was useful for tracing family trees because it's inherited only from the mother. It's not a mixture of both parents' genes, like nuclear DNA, so it preserves a family record that isn't scrambled every generation. It's altered only by mutations—random, isolated mistakes in copying the genetic code, which are then passed on to the next generation. Each random mutation produces a new type of DNA as distinctive as a fingerprint. (The odds against two identical mitochondrial DNA's appearing by chance are astronomical because there are so many ways to rearrange the units of the genetic code.)

To study these mutations, the Berkeley researchers cut each sample of DNA into segments that could be compared with the DNA of other babies. The differences were clear but surprisingly small. There weren't even telltale distinctions between races. "We're a young species, and there are really very few genetic differences among cultures," Stoneking says. "In terms of our mitochondrial DNA, we're much more closely related than almost any other vertebrate or mammalian species. You find New Guineans whose DNA is closer to other Asians' than to other New Guineans'." This may seem odd, given obvious racial differences. In fact, though, many differences represent trivial changes. Skin color, for instance, is a minor adaptation to climate — black in Africa for protection from the sun, white in Europe to absorb ultraviolet radiation that helps produce vitamin D. It takes only a few thousand years of evolution for skin color to change. The important changes — in brain size, for instance — can take hundreds of thousands of years.

The babies' DNA seemed to form a family tree rooted in Africa. The DNA fell into two general categories, one found only in some babies of recent African descent, and a second found in everyone else and the other Africans. There was more diversity among the exclusively African group's DNA, suggesting that it had accumulated more mutations because it had been around longer — and thus was the longest branch of the family tree. Apparently the DNA tree began in Africa, and then at some point a group of Africans emigrated, splitting off to form a second branch of DNA and carrying it to the rest of the world.

All the babies' DNA could be traced back, ultimately, to one woman. In itself that wasn't surprising, at least not to statisticians familiar with the quirks of genetic inheritance. "There *must* be one lucky mother," Wilson says. "I worry about the term 'Eve' a little bit because of the implication that in her generation there were only two people. We are not saying that. We're saying that in her generation there was some unknown number of men and women, probably a fairly large number, maybe a few thousand." Many of these other women presumably are also our ancestors, because their nuclear genes would have been passed along to sons and daughters and eventually would have reached us. But at some point these other women's mitochondrial genes disappeared because their descendants failed to have daughters, and so the mitochondrial DNA wasn't passed along. At first glance it may seem inconceivable that the source of all mitochondrial DNA was a single woman, but it's a well-established outcome of the laws of probability.

You can get a feel for the mathematics by considering a similar phenomenon: the disappearance of family names. Like mitochondrial DNA, these are generally passed along by only one sex — in this case, male. If a son marries and has two children, there's a one-in-four chance that he'll have two daughters. There's also a chance that he won't have any children. Eventually the odds catch up and a generation passes without a male heir, and the name disappears. "It's an inevitable consequence of reproduction," says John Avise, a geneticist at the University of Georgia. "Lineages will be going extinct all the time." After 20 generations, for instance, it's statistically likely that only 90 out of 100 original surnames will disappear. Avise cites the history of Pitcairn Island in the Pacific, which was settled in 1790 by 13 Tahitian women and six British sailors who had mutinied on the *Bounty*. After just seven generations, half of the original names have disappeared. If the island remained isolated, eventually everyone would have the same last name. At that point a visitor could conclude that every inhabitant descended from one man—call him the Pitcairn Adam.

So thus there must be a mitochondrial Eve, and even traditional anthropologists can't really argue against her existence. What shocked them about Mitochondrial Mom was her birthday, which the Berkeley researchers calculated by counting the mutations that have occurred to her DNA. They looked at the most distant branches of the family tree— the DNA types most different from one another—and worked backward to figure out how many steps it would have taken for Eve's original DNA to mutate into these different types. They assumed that these mutations occurred at a regular rate—a controversial assumption that might be wrong, but which has been supported by some studies of humans and animals. Over the course of a million years, it appears that 2 to 4 percent of the mitochondrial DNA components will mutate. By this molecular calculus, Eve must have lived about 200,000 years ago (the range is between 140,000 and 290,000 years). This date, published this past January by the Berkeley group, agrees with the estimate of a team of geneticists led by Douglas Wallace of Emory University.

But the Emory researchers think Eve might have lived in Asia. They base their conclusion also on mitochondrial DNA, which they gathered from the blood of about 700 people on four continents. They used different methods in chopping up the DNA and arranging the types in a family tree. Their tree also goes back to one woman, who lived 150,000 to 200,000 years ago, they estimate. Unlike the Berkeley researchers, however, they found that the races have distinctive types of DNA. They also found that the human DNA type most similar to that of apes occurred at the highest frequency in Asia, making that the likely root of the family tree. Wallace's data suggests that Eve can be traced to southeast China, but he cautions that this is only one possible interpretation of the data. "If we make other assumptions, we can run our data through a computer and come up with a family tree starting in Africa," he says. "So I'm not ruling out Africa. I'm just saying that we can't yet decide whether it's Asia or Africa."

The rival geneticists are quick to criticize one another. Wallace faults the Berkeley researchers for getting most of their African DNA samples from American blacks, whose ancestors could have mixed with Europeans and American Indians. The Berkeley researchers insist that their study is better because they chopped the DNA into smaller pieces, enabling them to analyze differences more carefully. Both groups acknowledge that there's room for improvement, and they're planning to gather more samples and look more closely at the DNA's structure.

At the moment, the evidence seems to favor an African Eve, because other genetic studies (of nuclear DNA) also point to an origin there and because that's where the earliest fossils of modern humans have been found. But wherever Eve's home was, the rival geneticists agree that she lived relatively recently, and this is what provokes anthropologists to start arguing—often with Biblical metaphors of their own.

If Eve lived within the past 200,000 years, she may have been a modern human, perhaps one of the first to appear. In that case she might have looked like a more muscular version of today's Africans. Or maybe it was her descendants who evolved into modern humans. Eve herself might have been our immediate ancestor, an archaic Homo sapiens, and therefore brawnier, with a large, protruding face and a forehead receding behind prominent brow ridges. She was certainly a hunter-gatherer, probably much like today's Bushmen in southern Africa, living in a group of maybe 25, carrying a nursing child across the plains in search of food. Humans around the world—Java man, Peking man— were living like this for hundreds of thousands of years before our mitochondrial Eve.

The question is: what happened to all the other populations around the world? For their women's mitochondrial genes apparently all vanished. The Berkeley biologists conclude that everyone outside Africa stems from a group of Eve's descendants who left their homeland between 90,000 and 180,000 years ago. As they moved across Asia and Europe, they would have encountered Neanderthals and populations of archaic Homo sapiens. They were probably outnumbered in many places. But wherever the daughters of Eve went, only their mitochondrial DNA survived.

Did the immigrants kill the natives? Possibly, but the conquests may have been peaceful. Because they were modern humans, Eve's descendants were less muscular than the archaic natives, but they were more organized, more able to plan ahead. They could make better stone tools. As they prospered and multiplied, consuming more of the local fruit and game, the natives would have suffered; a slight increase in their mortality rate could have led to their extinction in just a thousand years.

The immigrants may have been able to interbreed with the locals. Some anthropologists see physical vestiges of the Neanderthals in modern Europeans, and the Eve hypothesis doesn't rule out the possibility that the Neanderthals' nuclear genes were passed along to us. But the fact remains that the Neanderthals' mitochondrial genes all disappeared after Eve's descendants arrived, so both the Berkeley and Emory biologists suspect there was little or no mixing. Maybe the immigrants were so different that they couldn't interbreed. Or maybe they simply shunned the natives as being too "primitive." The Neanderthals' attempts at courtship presumably suffered if, as some scientists speculate, they lacked modern humans' power of speech.

This question of interbreeding is the crux of the bones–molecules debate. The geneticists' most vehement critic is Milford Wolpoff, a University of Michigan paleoanthropologist who believes our common ancestor lived closer to a million years ago. "The most obvious conclusion from the genetic evidence," he says, "is that Eve's descendants spread out of Africa and weren't incorporated at all into the local populations. I find that incredible. In recorded history, there always has been intermixing as populations moved or villages exchanged wives. I believe we have a long history of people constantly mixing with one another and cooperating with one another and evolving into one great family." Wolpoff finds his version of evolution more satisfying than "this business about Eve showing the common nature of everything." If Eve's descendants wiped out all rivals, Wolpoff suggests, maybe the theory should be named after her murderous son, Cain.

Actually, the more common term for this idea is Noah's Ark, coined by Harvard's W. W. Howells in describing the two classic schools of anthropological thought on the origin of modern humans. One school believes that a small group of modern humans appeared in one place recently—perhaps 100,000 to 200,000 years ago—and colonized the entire world, like the survivors of Noah's ark. The other populations were not inexorably climbing the ladder or the tree of evolution—they were more like twigs on a bush or the arms of a hatrack, branching off to an inglorious end. The idea of a recent common origin for humanity was held by many anthropologists long before DNA provided supporting evidence.

The opposing school believes in what Howells called the "candelabra hypothesis": the different races diverged long ago and evolved independently into modern humans, progressing like the parallel candles of candelabra. This view became prominent in 1962 with Carleton Coon's book *The Origin of Races*. He insisted that modern humans did not suddenly appear, "fully formed as from the brow of Zeus," in one place. "I could see that the visible and invisible differences between living races could be explained only in terms of history," wrote Coon, a University of Pennsylvania anthropologist. "Each major race had followed a pathway of its own through the labyrinth of time."

Unfortunately, Coon published his theory along with a speculation that was denounced as racist. He suggested that African civilization was less advanced because black people were the last to evolve into modern humans. Although the first hominids may have arisen in Africa, Coon said, the evolution of modern humans seemed to occur first in Europe and Asia. "If Africa was the cradle of mankind, it was only an indifferent kindergarten." He couldn't have been more wrong. Bones subsequently discovered in Africa are believed to be from modern humans living there about 100,000 years ago. These bones (as well as some from Israel that might be as old) represent the earliest known modern humans. Before their discovery it was assumed that modern humans didn't evolve until 35,000 years ago, which is when they first appear in the European fossil records. So blacks were hardly the last to reach modernity.

Coon's mistake didn't invalidate the basic candelabra hypothesis, which is still popular in a modified version. Wolpoff prefers to think of a trellis: the separate races gradually evolving along parallel lines but connected by a network of genes flowing back and forth. The Neanderthals turned into modern Europeans while Peking man's descendants were becoming modern Chinese. Immigrants brought in new genes, but the natives' basic traits survived. This would explain why both the Neanderthals and the modern Europeans have big noses, why Peking man and current residents of Beijing have flat faces, why today's Aboriginal Australians have flat foreheads like Java man. These similarities presumably wouldn't persist if the ancient natives had disappeared when Eve's descendants arrived.

Other anthropologists, however, find these similarities unconvincing. It might just be a coincidence that modern Europeans have big noses like the Neanderthals. To the Noah's–ark school, what's striking are the *differences* between ancients and moderns. Modern Europeans, for instance, are much less stocky than Neanderthals—their arms and legs are proportioned more like those of humans from the tropics, as Eve's presumably were. And there's no clear sign in the fossil records of a transition from Neanderthal to modern. Some anthropologists cite bones that might belong to hybrids of immigrants and natives, but these interpretations are disputed.

"I don't rule out the possibility that there was interbreeding, but I don't see it in the fossils," says The British Museum's Christopher Stringer. "In the two areas (where) we have the best fossil evidence, Europe and Southwest Asia, the gap between archaic and modern people is very large. The entire skeleton and brain case changed. I think the fossil evidence is clearly signaling replacement of the archaic population. I was delighted to see the DNA results support this view."

Most anthropologists, though, are still skeptical. They don't reject outright the genetic evidence, but they don't accept it flatly, either. After the mistakes of the past, they're leery of any grand new theory about human evolution. They rightly point out that the geneticists' molecular clock could be way off—change a few assumptions and Eve's birthday could move back hundreds of thousands of years, bringing ancients like Peking man back into our lineage. Above all, anthropologists would like to see the corroborating bones.

"We don't know what's going on here," says the University of Pennsylvania's Alan Mann. "Maybe we are dealing with a dramatic jump. Maybe the origin of creatures like us occurred very recently. Certainly the mitochondrial data is a significant advance. But there really isn't any good fossil evidence from that period to back it up. If you look at the fossils, the good evidence on Africa can be placed on the palm of your hand. In this field, a person kicks over a stone in Africa, and we have to rewrite the textbooks."

So the fossil hunters will keep digging—now they have something specific to look for in the sediments of 200,000 years ago. Maybe they'll vindicate the geneticists once again, but the geneticists aren't waiting to find out. They're already trying to expand the Eve theory by finding Adam. Researchers in England, France and the United States have begun looking at the Y chromosome, which is passed along only on the male side. Tracing it is difficult because it's part of the DNA in the cell's nucleus, where there are many more genes than in the mitochondrion. This Adam will be the one lucky father whose descendants always had at least one son every generation. He may have been hunting and gathering while Eve was, or he may have lived at another time (though it would cast doubt on the Eve hypothesis if the time and place of his birth were too distant). The researchers hope to get an answer within several years.

In the meantime, there is one temporary candidate for an Adam—not the one scientists are looking for but one defined simply as a man from whom we are all descended. Since we are all descended from Eve's daughters, any common male ancestor of theirs would be a common ancestor to everyone today. This wouldn't necessarily be Eve's husband. For all we know, she may have had more than one. But her daughters all certainly had the same maternal grandfather. So, at least for now, the only safe conclusion is that Adam was Eve's father.

How would you describe Tierney's writing style? What approaches does Tierney use to fairly present the ideas of geneticists and anthropologists? Why is the tracer inherited only on the mother's side? Why did Tierney wait until two-thirds of the way through the article to explain this?

Exercises

1. You're probably familiar with the opening of Abraham Lincoln's Gettysburg address:

   > Four score and seven years ago, our fathers brought forth on this continent a new nation conceived in liberty and dedicated to the proposition that all men are created equal.

   But this, according to Oliver Jensen, is what the Gettysburg Address would have sounded like if it had been written by Dwight Eisenhower:

   > I haven't checked these figures but around 87 years ago, I think it was, a number of individuals organized a governmental set-up here in this country, I believe it concerned the Eastern states, with this idea they were following up based on a sort of national independence arrangement and the program that every individual is just as good as every other individual.

   Jensen's parody doesn't say anything more than Lincoln's original, yet it is much less convincing. Had Lincoln used Eisenhower's text at Gettysburg, do you think anyone would have remembered what he'd said? Why? What are the consequences of such writing?

2. Economist John Kenneth Galbraith argues the case against "elitist" writing:

   > There are no important propositions that cannot be stated in plain language. The writer who seeks to be intelligible needs to be right; he must be challenged if his argument leads to an erroneous conclusion and especially if it leads to the wrong action. But he can safely dismiss the charge that he has made the subject too easy. The truth is not difficult. Complexity and obscurity have professional value—they are the academic equivalents of apprenticeship rules in the building trades. They exclude the outsiders, keep down the competition, preserve the image of a privileged or priestly class. The man who makes things clear is a scab. He is criticized less for his clarity than for his treachery.

   Use evidence to support or attack Galbraith's argument.

3. Edit this sentence from a Nuclear Regulatory Commission report:

   > It would be prudent to consider expeditiously the provision of instrumentation that would provide an unambiguous indication of the level of fluid in the reactor vessel.

4. Edit this sentence from a sign posted at an entrance to a corridor in the Pentagon:

   > This Passageway has been made nonconductive to utilization for an indefinite period.

5. Write a paragraph about the issues raised in this quotation by P. Meredith:

   > The great heresy of science has been its adoption of impersonal language. For every observation involves a private world, and science is based on observation . . . we see, then, that to restore the personal pronoun to the language of science is much more than a literary facilitation of readability. It

demands the self-understanding of science as an arbitrary enterprise of man. Too long it has masqueraded as the impersonal voice of Nature, another God issuing commandments. And by forcing the restoration we shall force the examination of the values which dominate our choice of axioms. Moreover, this is no mere verbal problem, but a private evaluation of our public human relations. Scientists must restore their own atrophied feelings.

6. Edit this sentence to improve its clarity and readability:

   The results of the present study suggest that in addition to the manifestation of aberrant homeostatic patterns of neurohumoral activity following the cessation of noxious stimulation, the neurotic may be further characterized by atypical autonomic responses to an increase in the level of appetitional drives.

7. Edit this paragraph to improve its clarity and readability:

   This small monograph is an excellent summary of current concepts of animal physiology. It is profusely illustrated with beautiful pictures and clear drawings. The text is relatively simple and is obviously written for the non-expert, for there are very few references cited.

8. A famous writer once claimed the "passive voice doesn't look you in the eye." What did she mean by this?

9. In 1967 F. Peter Woodford described in delightful detail the poor writing in "journals—even the journals with the highest standards":

   In the linked worlds of experimental science, scientific editing, and science communication many scientists are considering just how serious an effect the bad writing in our journals will have on the future of science.

   All are agreed that the articles in our journals—even the journals with the highest standards—are, by and large, poorly written. Some of the worst are produced by the kind of author who consciously pretends to a "scientific scholarly" style. He takes what should be lively, inspiring, and beautiful and, in an attempt to make it seem dignified, chokes it to death with stately abstract nouns; next, in the name of scientific impartiality, he fits it with a complete set of passive constructions to drain away any remaining life's blood or excitement; then he embalms the remains in molasses of polysyllables, wraps the corpse in an impenetrable veil of vogue words, and buries the stiff old mummy with such pomp and circumstance in the most distinguished journal that will take it. Considered either as a piece of scholarly work or as a vehicle for communication, the product is appalling.[15]

   Do you think this criticism is true today? Provide examples to support your answer.

10. Make these sentences and paragraphs more concise:

    The university deemed it necessary to terminate the professor.

    Model: *The university fired the professor.*

    It is utterly fruitless to become lachrymose over precipitately decanted lacteal fluid.

    Model: *Don't cry over spilled milk.*

We could, of course, proceed on the assumption that an animal's behavior is innate.

I want to be apprised continuously of your progress.

Do not work on the electron microscope when smoking.

The desirability of abbreviating the incubation period was worth mentioning by the professor.

Assessment of the experiment should precede implementation of the program.

Randy's data failed to impact positively on me.

Discussions with fellow students in other classrooms elucidated that the problem of poor writing is not uncommon.

Another very and extremely important consequence of Watson and Crick's model is that, at this point in time, we now have a testable hypothesis for further consideration and experimental documentation.

The scope and magnitude of tasks that need to be performed in order to quantitatively measure the precise ultrastructure has resulted in the publication of a manual at this point in time by us.

Microscopic examination of the microscope slide was accomplished so you could see the degree to which the tissue is stained.

The purpose of this experiment is to offer the student the ways and resources to assist those in charge of developing protocols and procedures to increase the effective utilization of the amino acid analysis information system.

It is concluded that our current experiments that are now underway will greatly simplify our current model for calcium transport in epithelial cells.

A surgically implanted device devised to normalize the rhythm of a patient's heart and greatly reduce the death of at-risk cardiac patients is called an implantable automatic defibrillator.

The *Biology Club* at my educational institution is an organization of students dedicated and committed to development of further improvement in writing to learn biology so as to enhance the expected writing skills for the beginning biologist. Application forms for admission are obtainable for any member.

It is the opinion of this writer that it is the appropriate time and place to reexamine the mode of writing which might most effectively be utilized by members of the biological profession. It is also the writer's firm belief that the inappropriateness of the active voice and the personal pronoun for scientific writing has truly made for a great deal of inefficiency regarding writing. This nonpersonal style of writing has numerous difficulties and other problems that should be eliminated at the earliest possible date.

[15]Woodford, F. Peter. 1967. Sounder thinking through clearer writing. *Science* 156: 743.

Health educators serve as change agents: they facilitate lifestyle alterations in recipients of their services.

Feathered bipeds of similar plumage will live gregariously.

A revolving volcanic conglomeration accumulates no congeries of small bryophytic herbage.

The temperature of the aqueous content on an unremittingly ogled vessel will never attain 212 degrees Fahrenheit.

Abstention from uncertain undertakings precludes a potential escalation of remuneration.

Ingestion of an apple (the pome fruit of any tree of the genus, *Malus*, said fruit being usually round in shape and red in color) on a diurnal basis will with absolute certainty keep a primary member of the disease prevention establishment absent from one's local environment.

Scintillate, scintillate, astra minific.

Neophytic serendipity.

Sorting by mendicants must be extirpated.

As a case in point, other authorities have proposed that slumbering canines are best left in a recumbent position.

Circumspection requires that individuals abiding in vitreous edifices refrain from launching petrous projectiles.

There is no utility in belaboring a deceased equine.

I do not want to read this obsolete operator's manual.

The experiment was delayed by the ongoing critical lab animal shortage problem.

Hoping for a better crop, the field was irrigated weekly.

Being confined to a small cage, the zoologist tried to enrich the baby baboon's food.

While looking through the microscope, a new idea popped into my head.

Since hiring the technician, our data have improved.

Reaching the heart, a bypass was performed on the blocked arteries.

At the age of 12 my mother entered me in a science fair.

To become a biology professor, two degrees must be earned.

Spending four hours on the operating table, a tumor the size of a grapefruit was removed.

Taking a deep breath, the incision was started.

To learn the technique the first three exercises must be completed.

As an experienced researcher, the problem caused little trouble.

To increase their research, three new rats were purchased.

I looked for the book about poisonous fungi on the shelf in my office.

The professor spoke to the student with a harsh voice.

Ling-Ling was watched eagerly as she produced the first live baby panda by the friends of the National Zoo.

He diagrammed the new model for cellular communication on the pads of paper on the way to Waco.

Give the petri dishes and desk to the student with skinny legs.

We will study in the library of a biologist of great quality.

The drug is highly effective in suppressing intestinal flu. It is unfortunate that it is not prescribed more often.

The American crocodile could once be found throughout southern Florida. It is now threatened with extinction.

The biologist told Jim that he had the petri dishes.

He told the professor his pH meter was broken.

Martha told Jennifer that she was worried about her data.

The bottom line is that Darwin was right.

A new microscope will benefit each and every one of us if we only give it 110%.

The first experiment went off without a hitch.

My lab partner is as nutty as a fruitcake.

Failure of the treatment with penicillin could not have been unpredicted because of the defective assay method used. Unfortunately, this occurs in many clinics.

Jimmy was unsuccessful in his attempt to pass the test.

Plans for the field survey will be made by the students.

The need for better microscopes exists in our laboratory.

Visual observation was made of the microscope slides that were utilized which were stained with stain.

The absorption coefficient of the chlorophyll must be determined over the sample range by the students in the classroom.

Pollution constitutes a threat to wildlife.

All data were subjected to standard statistical analyses for verification.

Attempts to isolate the virus from the bacterium gave negative results.

The students reached a decision to terminate the experiment after extensive deliberations.

The biology students will conduct an investigation of the data.

The new computer is quieter, faster, and will break down less often than the current computer.

These data provide support for your conclusions.

Weekly updates can be expected hereafter.

It was demonstrated that the cells contain lipid.

The paper was short, logical, and reading it was easy.

They cannot do a verification of the data until you make a decision about the equipment.

This paper furnished no explanation of your methods.

We can carry out the examination of animals during the next two week period.

The experiment will be done by the graduate students.

The decision was made to repeat the experiment.

The experiment will be done by me.

A series of standards, varying in concentration over the range corresponding to approximately 0.5 to 4 times the standard, is prepared and analyzed under the same conditions and during the same period of time as the unknown samples.

DDT, if it comes into contact with the skin, if it is used in areas where inhalation is possible, is subject to severe restrictions.

Future planning favors termination of access to the electron microscope facility during non-classtime hours.

Included in these plans is the construction of a high quality, multiuser, easy access, biotechnical center for optimization of research output.

Recycling of the pre-used inventory has had a positive effect on the visual and economic status of the area immediately adjacent to the university.

If the measurement instructions described in the manufacturer's users brochure are followed, then there will be a small margin for error in our results.

Before analysis of the samples, the analytical system is allowed to equilibrate until a steady baseline data reading is attained and reached on the recorder.

Any chemical substance present in the growth medium at a high enough level can interfere with the contiguous growth of the bacteria.

Improvements in quality control have been effected in our new labs in the basement.

We finally reached a decision calling for the substitution of calcium for magnesium in our media.

We made a careful consideration of costs before reaching the decision to begin our experiments.

The replacement of a lens is a job that can be handled by the least experienced student.

In some diseases, the liver produces an excessive amount of porphobilinogen, an inborn error of metabolism.

The reason why the experiment failed is because the wrong temperature was used.

If an identification of minor flaws in our experimental procedure can be made before next week, there will be fewer problems.

In hypotonic constipation the colon is usually partially full from one end to another.

Errors were discovered in the calculation.

The end of a sentence is the place where the most important information should be placed.

A new genus of earthworms was discovered by a biology student.

The results were misplaced by Wally.

This chapter serves the function of discussing the main families of plants.

It can be concluded that my data are worthless.

Diseases have been discovered more often than cures by biologists.

It is a requirement that a copy of the completed report be sent to the professor as soon after termination of the lab as is reasonably feasible.

Research goals will be established by the National Science Foundation.

It was suggested by the students that lab reports be accepted late.

In the next chapter the functions of the nucleus are explained.

Compilation of the data should be completed by the students no later than 24 December 1991.

11. Discuss, support, or refute the ideas contained in these quotations:

Plants, instead of affecting the air in the same manner as animal respiration, reverse the effect of breathing and tend to keep the atmosphere sweet and wholesome. — Joseph Priestly

The greatest service which can be rendered any country is to add a useful plant to its culture; especially, a bread grain; next in value to bread is oil. — Thomas Jefferson

It is not so much that the cells make the plant; it is rather that the plant makes the cells. — Heinrich Anton de Bary

The structure of tissues and their functions, are two aspects of the same thing. — Alexis Carroll

12. Write a short essay describing how this *The Far Side* cartoon relates to biology:

THE FAR SIDE   By GARY LARSON

Great moments in evolution.

Figure 4–1
The Far Side. Gary Larson © 1982 Universal Press Syndicate

CHAPTER FIVE

Cohesion

Science is built up with facts, as a house is with stones. But a collection of facts is no more a science than a heap of stones is a house.
   —Henri Poincaré

If your research is strong, then everyone should know it. So state what you did up front—the first paragraph, even the first sentence—and leave the dilly-dallying for people who don't have anything to report.
   —Erich Kuhnardt

In whatever paragraphs or essays you write, verify the sequence of ideas and take out or transpose everything that interrupts the march of thought and feeling.
   —Jacques Barzun, *Simple and Direct*

The best argument is that which seems merely an explanation.
   —Dale Carnegie

So far I've discussed clarity as if biologists write only individual sentences. Simple, concise, and precise sentences increase communication but can also produce dull prose—noble and virtuous, but dull. Matching characters and actions to subjects and verbs creates clarity, but this clarity is a local clarity restricted to only one sentence. This is another instance where mere rules such as "Be clear" will fail you; indeed, readers may understand all of your words and sentences but may not know what you're trying to say. Since clear sentences can confuse readers if the sentences are not in the proper context, good writing requires more than accuracy and local clarity. Writing an effective paragraph requires a continual compromise between local clarity and cohesion, with priority given to the cohesiveness that helps readers see the paragraph's message.

The most important concerns of a writer are not individual topics of individual sentences but the cumulative effect—the communication—of a cohesive sequence of sentences. These sequences of sentences are paragraphs, the major units of scientific writing. Paragraphs are logically constructed passages having one theme. Concise, effective sentences are a must, because if one sentence is weak, language falters and the reader stumbles.

The cliché stressed in most writing books to cover cohesion is, "Each paragraph should communicate one idea." Yes, a paragraph *should* communicate an idea. But how do you ensure that? Since it takes more than one sentence to describe most ideas, implementing this principle requires that a writer know how to arrange sentences into an effective paragraph. This involves understanding how readers read sentences—not individually, but in sequences.

Readers divide paragraphs into two parts: (1) a summary of the paragraph's take-home message and (2) points that stress the take-home message. The take-home message is best placed in a short opening sentence, with the stress toward the end of the sentence. The rest of the paragraph is where the writer develops, and where readers look for, new ideas. Notice how the first sentence of this paragraph introduces evolution, while later sentences develop the idea:

> Clark's practice of carefully mapping every fossil made it possible to follow the evolutionary development of various types through time. Beautiful sequences of antelopes, giraffes and elephants were obtained; new species evolving out of old and appearing in younger strata. In short, evolution was taking place before the eyes of the Omo surveyors, and they could time it. The finest examples of this process were in several lines of pigs which had been common at Omo and had developed rapidly. Unsnarling the pig story was turned over to paleontologist Basil Cook. He produced family trees for pigs whose various types were so accurately dated that pigs themselves became measuring sticks that could be applied to fossils of questionable age in other places that had similar pigs.[1]

[1]Johanson, Donald C., and Maitland A. Edey. 1981. *Lucy: The Beginnings of Humankind.* New York: Simon and Schuster. pp. 116–117.

Chores on the Word Farm

After reading all of these suggestions for improving punctuation, grammar, and style, you may want a few simple guidelines to help with your writing. Almost 50 years ago, George Orwell provided some advice that I can't improve:

> Never use a metaphor, simile, or other figure of speech which you are used to seeing in print.

> Never use a long word where a short one will do.

> If it is possible to cut a word out, always cut it out.

> Never use the passive voice where you can use the active [voice].

> Never use a foreign phrase, a scientific word or a jargon word if you can think of an everyday English equivalent.

> Break any of these rules sooner than say anything outright barbarous.

For more advice from George Orwell, see pages 116 and 126.

Although effective paragraphs are made of simple, straightforward sentences, a paragraph is much more than a group of related sentences. Rather, a well-written paragraph shows relationships and makes a point. To show relationships you must link the ideas of the paragraph. You'll do this most effectively by (1) organizing sentences in a logical order and (2) linking these sentences with transitions that lead readers to your conclusion.

Organize Sentences in a Logical Order

The cohesion of a paragraph depends on the arrangement of concise sentences. This arrangement must reflect the logic behind your argument and develop your idea. Here are some suggestions for writing effective paragraphs:

The most prominent and important part of a paragraph is its first sentence. Use the first sentence to tell readers what the paragraph is about. Don't make readers be detectives; tell them the significance of your work. Use your first sentence or paragraph to tell readers that the butler did it, and the rest of the paragraph or paper to describe how it happened. Again note the opening sentence of Watson's and Crick's monumental paper from *Nature*:

> We wish to suggest a structure for the salt of deoxyribose nucleic acid (D.N.A.).

Develop the paragraph's idea with data, examples, comparisons, and analogies. Watson and Crick drove home the importance of their discovery with a great understatement:

This structure has novel features which are of considerable biological interest.

Delete sentences irrelevant to the paragraph's topic sentence. By doing so, you'll stick to the point of the paragraph. Notice how Erik Denton does this by starting his article with an interesting comparison:

Many fishes, like many people, must keep swimming just to keep from sinking. Bone and muscle are denser than sea water and so tend to drag the animal to the bottom. It is obvious that the ability to float can be advantageous for animals that live in the sea. Accordingly, fishes are equipped with swim bladders which give them neutral buoyance—that is, an average density equal to that of sea water—and save them the labor of continuous swimming. Two other animals—the cuttlefish and cranschid squid—have developed quite different kinds of flotation organs. They anticipated man in using the working principles of the submarine and the bathyscaph, the one endowing the cuttlefish with active control of its buoyancy, the other permitting the squid to hover at greater depths.[2]

Link each sentence to the ones preceding and following so that they produce a coherent idea. In the following paragraph from *The Origin of Species*, Charles Darwin's logic dictates the order of sentences. He expresses some thoughts with simple sentences or independent clauses, while expressing others as dependent clauses. Finally, he uses transitions to show relations between the sentences and clauses, concessions and exceptions, correlations and disjunctions:

The truth of the principle, that the greatest amount of life can be supported by great diversification of structure, is seen under many natural circumstances. In an extremely small area, especially if freely open to immigration, and where the contest between individual and individual must be severe, we always find great diversity in its inhabitants. For instance, I found that a piece of turf, three feet by four in size, which had been exposed for many years to exactly the same conditions, supported twenty species of plants, and these belonged to eighteen genera and to eight orders, which shows how much these plants differed from each other. So it is with the plants and insects on small and uniform islets; and so in small ponds of fresh water. Farmers find that they can raise most food by a rotation of plants belonging to the most different orders: nature follows what may be called a simultaneous rotation. Most of the animals and plants which live close round any small piece of ground, could live on it (supposing it not to be in any way peculiar in its nature), and may be said to be striving to the utmost to live there; but, it is seen, that

---

[2]Denton, Erik. 1960. "The Buoyancy of Marine Animals." *Scientific American*, July, 1960.

where they come into the closest competition with each other, the advantages of diversification of structure, with the accompanying differences of habitat and constitution, determine that the inhabitants, which thus jostle each other most closely, shall, as a general rule, belong to what we call different genera and orders.

Writing is clear when readers can quickly get your message and follow your supporting arguments. Anything else is bad writing. Therefore, write so that readers get your message quickly and on the first reading. Don't hesitate to use comparisons to help readers visualize your ideas. For example:

A virus is to a person as a person is to the earth. Likewise, the size of a human cell is to that of a person as a person's size is to that of Rhode Island; an atom is to a person as a person is to the earth's orbit around the sun; and a proton is to a person as a person is to the distance of Alpha Centauri.

End the paragraph with a short, direct, concluding sentence, and start a new paragraph to announce a change of subject. For example, in *In the Shadow of Man*, Jane Goodall describes a baby chimpanzee named Pom, who, unlike other baby chimps, always stayed close to her mother. She closes the chapter's last paragraph with this emphatic sentence: "Obviously she was terrified of being left behind."

Start a new paragraph when you start describing a new idea. This usually corresponds to places where you would say, "Oh, by the way . . ." or "Here's something else . . ."

Remember that readers grasp material most readily when it is presented in units of 75 to 200 words (5 to 12 sentences). Avoid writing paragraphs longer than a page or so.

If your text is more than about four paragraphs long, consider using headings and subheadings to divide the text. Use these subheadings to tell readers what's ahead.

Writers must communicate with their readers. To communicate, a writer must show, and to show, a writer must provide details. General statements provide an overview, but they alone are unconvincing. For example, a general statement like, "Microbes are interesting" only prompts readers to ask for details — *how* are microbes interesting? Examples and details give meaning to generalizations. Readers remember details and examples: they want to hear about miles per gallon, not that a car "runs good." Therefore, instead of saying that a piece of apple pie contains "a lot of energy," write that it "contains about 1.5×10^6 J (350 Cal) of energy, enough energy for a woman to run for almost an hour or for a typist to enter about 15,000,000 characters on a manual typewriter."

Be as specific as possible. Just as you are annoyed by vague sentences such as, "Dr. DeBakey inserted a valve in the vicinity of the heart," so too will readers be annoyed if you do not give them enough details about your subject. Notice how these generalizations improve when you add details:

Bioenergetics
GENERAL: A piece of apple pie contains a lot of energy.
SPECIFIC: The energy in a piece of apple pie (350 Calories; 1.5×10^6 J) equals the energy content of 11.2 ounces of TNT and is enough energy to fuel a one-hour run by a woman. The 3,000 calories of food energy needed daily by most college students are contained in about 9 pieces of apple pie and are equivalent to the energy in 6 pounds of TNT.

Humans
GENERAL: Fetuses grow fast just before birth.
SPECIFIC: Fetuses grow fastest in the last three months before birth. If infants continued to grow at that rate, they would be 18' 4" tall when they reached age 10.

Mammals
GENERAL: Some prehistoric horses were small.
SPECIFIC: *Eohippus*, a prehistoric horse, was about the same size as a modern cat.

Dinosaurs
GENERAL: Dinosaurs were large but had small brains.
SPECIFIC: The *stegosaurus*, a playful 25'-long dinosaur, had a brain the size of a walnut.

Marine creatures
GENERAL: Some sponges are small.
SPECIFIC: It would take a long time to wash with *Leucosolenia blanca*, the smallest known sponge. When fully grown, more than 30 of these sponges would fit end-to-end across the palm of your hand.

Microbes
GENERAL: Some bacteria form interesting shapes.
SPECIFIC: *Chondromyces* bacteria climb atop each other to form towers several millimeters high. On a human scale, these towers would be over a mile high, or some four times the height of the World Trade Center in New York City.

Plants
GENERAL: Trees can help homeowners save energy.
SPECIFIC: A mature tree in front of your home can produce a cooling effect equal to that of 10 room-size air-conditioners running 20 hours per day.

Rules of Writing

Never, ever use repetitive redundancies.

Never use a long word when a diminutive one will do.

Use parallel structure when you write and in speaking.

Place pronouns as close as possible, especially in long sentences—such as those of 10 or more words—to their antecedents.

Proofread your paper carefully to find any words that you out.

Do not let a colon separate: the main parts of a sentence.

Avoid overuse, of commas.

Delete unnecessary, excess words that are not needed.

Avoid run-on sentences they are hard to read.

Put an apostrophe where its needed.

Do not put statements in negative form.

Do not use a hyphen when it is un-necessary.

Do not use a semicolon where; it is not needed.

Don't abbrev.

Have a good reason for Capitalizing a word.

Consult a dictionary for prosper spelling.

After studying these rules, dangling modifiers will be easy to correct.

Avoid using trendy words whose parameters are not viable.

Remember that verbs has to agree with their subjects.

Each pronoun should agree with their antecedent.

Don't use no double negatives.

Remember to never split an infinitive.

No sentence fragments. Enough of this.

Pollution
GENERAL: Plants can help solve the pollution problem.
SPECIFIC: It takes more than 100,000 trees to cancel the pollution of one jet flying round-trip from New York to Los Angeles.

Size relationships
GENERAL: Amoeba are larger than *Chlamydomonas* and bacteria.
SPECIFIC: If an amoeba were as big as an elephant, *Chlamydomonas* would be the size of a cat, and a bacterium would be the size of a flea.

Anchor general statements with details and examples. For example, write "eight" instead of "several," and instead of writing that "some salt was added to the medium," write that "I added 10.2 g of NaCl to the medium." Use numbers and exact words wherever possible.

We remember best what we hear first and last. This is why an effective paragraph needs a topic sentence summarizing its point and why an effective essay needs a summarizing sentence or paragraph to convey its message. This is also why most famous quotations are usually from the beginning or end of speeches or books. For example, "government *of* the people, *by* the people, *for* the people . . . [italics added]" comes from the end of Lincoln's *Gettysburg Address*, and "To be or not to be" opens one of Shakespeare's works. Similarly, consider the opening line of Dickens's *A Tale of Two Cities:* "It was the best of times, it was the worst of times. . . ." Most people have not read *A Tale of Two Cities* yet are familiar with its opening line. Indeed, since being published in 1859, this line is now overused to the point of being a cliché. Moreover, you'd be hard pressed to find anyone who can quote any other lines from the 600-page book.[3]

Vary the Length of Your Sentences

Even when you follow all of these suggestions, you may end up with a monotonous paragraph. Such monotony results not from the writing being "incorrect," but rather from all of the sentences being similar. Indeed, a string of short sentences makes writing choppy, and a string of long sentences makes a reader drowsy. To avoid this monotony and to communicate most effectively, write sentences short enough to be held in your reader's mind. Do this by building paragraphs with sentences having an *average* length less than 17 words—about the average sentence-length in *Newsweek* (the average length of sentences in most "scholarly journals" exceeds 25 words, while that of sentences in legal documents is 55 words).[4] However, don't make *all* your sentences less than 20

---

[3]The only other lines that someone might be able to quote from *A Tale of Two Cities* are these: "It is a far, far better thing that I do, than I have ever done; it is a far, far better rest that I go to, than I have ever known." Not surprisingly, these lines end the novel.

[4]Sentences of many popular (Ann Landers), famous (Winston Churchill), and inspiring (John Steinbeck) writers average about 15 words, as do sentences in magazines such as *Reader's Digest, Atlantic, Harper's*, and *Saturday Review*. Consider these analyses of *Wise Blood* (by Flannery O'Connor) and *Of Mice and Men* (by John Steinbeck):

	Wise Blood	*Of Mice and Men*
Average sentence length	15.3 words	15.2 words
Longest sentence	40 words	35 words
Shortest sentence	5 words	6 words

Average sentence-length usually increases with a writer's education. For example, sentences written by fourth graders average 11.1 words, those written by ninth graders average 17.3 words, and those written by college juniors average 21.5 words. Most college graduates (but ones lacking good training in effective writing) write sentences averaging about 25 words.

Mark Twain was a master of varying the length of sentences to produce rhythmic writing. Here's how he varied the number of words in the opening paragraph of *The Adventures of Huckleberry Finn:* 25, 14, 12, 3, 21, 33, 28, 8, 12, 35, 51, 16, 33, 4.

words long. If you do, you may communicate adequately, but you'll be like a singer who uses only one octave: You'll be able to carry the tune, but you'll be unable to produce much variety.

Vary the length of your sentences. Do this by introducing a topic with a short, direct sentence. Develop the topic with longer sentences, and conclude with short, direct sentences. Concentrate on the first and last words of sentences: These are the words that stick most with readers (see the example on page 141 about killer cells). Also remember that short sentences can carry tremendous punch; that's why they dominate good writing. Appreciate and use simple declarative sentences.

Make Smooth Transitions

Arranging sentences in what you feel is a logical order may still leave readers confused. This confusion results from your having written separate, albeit well-written, sentences rather than a cohesive paragraph. Consequently, each sentence appears independent of others rather than as a part of an argument. In short, you've not linked the sentences' ideas, and therefore your logic is hidden. When your transitions are either absent or too abrupt, each sentence yanks and surprises readers because they have no hint as to what's coming.

Walter Campbell has likened a good transition to a bridge, which "must be level, and alike at both ends." There are many ways to effectively link ideas. One way is to use a paragraph's last sentence to close one idea and introduce the next. A common way of doing this is to repeat an important word in the last sentence of one paragraph in the first sentence of the next. For example, notice how these writers use the words "testing" and "tests" to bridge these paragraphs:

> A company considers enrolling one of its brightest young female employees in an expensive training program. Before making the investment, it asks the woman to undergo biological testing with the latest diagnostic techniques, including genetic screening.
> The tests reveal an unexpected problem. . . .

Another way to link paragraphs involves ending one paragraph with a question and starting the following paragraph with the question's answer. For example, notice how Peter Raven uses a question, answer, and repeated idea to link these paragraphs in his article entitled "The Less-Noticed Worldwide Revolution."

> Many parents across the country reportedly have been stunned to discover that their teenagers are baffled about the significance of events in Eastern Europe and the Soviet Union over the past several months. How, these parents wonder, can young people be so ignorant about such historic occurrences?
> I share their concern, but I think many parents are as culpable as their children when it comes to awareness of another, equally important

upheaval taking place in the world today. I refer to the revolution in biology, which is likely to change the course of human history as profoundly as anything in today's political arena.

The recent transformation of the biological sciences . . .

Here's another example of linking paragraphs with a question and an answer:

National surveys assure us that public understanding of science has never been poorer while our national need for a scientific literate public has never been greater. We are faced with an endless array of issues—AIDS, pesticides, ecological Armageddon, space stations and Stealth bombers—that are inextricably entwined with science and technology. How are we to manage these issues if we do not understand them?

Scientists and science communicators must do more to enhance "science literacy" among their fellow Americans through books, television programs, museums and, perhaps most productively, our nation's schools and colleges.

You'll also link paragraphs effectively if you state an idea or observation in one paragraph and use the following paragraph to explain the significance of that idea or observation. For example:

Presidential elections and other dramatic news may capture the headlines, but one of the most profound events of our time has been the dramatic increase in life expectancy. Americans now live about 25 years longer than they did a century ago.

That is an essential fact to keep in mind when one considers whether it is ethical to carry out experiments on animals, a controversy that has existed for more than a century. The debate has become more charged over the past decade, with some radical animal-rights activists breaking into laboratories to "liberate" animals.

Notice also how this writer used "controversy" and "the debate" to link the sentences of the second paragraph. Here's another example:

Imagine you are an ambulance driver and someone's life depends on your finding an address—but you have no map. That's the dilemma faced by medical researchers, who must search for the cures to serious genetic diseases without knowing where the relevant genes are located on the human chromosomes.

The result is that research today proceeds much more slowly than it might on heart disease, cancer, certain kinds of Alzheimer's disease, cystic fibrosis and some 3,000 other disorders with a genetic component. . . .

Like the ambulance driver searching blindly along city streets, these researchers could do their work far more effectively—and save many more lives—if they had a decent road map.

An even simpler way is to bluntly tell readers what's coming. Here's how Charles Darwin did this:

> We will now discuss in a little more detail the struggle for existence.

You can also lead readers by linking the ideas of successive sentences with transitions that tell readers what to expect. For example, "therefore" spotlights a conclusion, while "however" and "but" warn readers that you're changing direction. Without these transitions, sentences remain independent, and readers have no time to grasp the full meaning of one idea before the next is thrust upon them.

Ironically, poor linkage of ideas is one of the most common symptoms of poor writing yet is one of the easiest problems to solve. Remember that *transitions connect related ideas* by showing relationships, by leading the reader to your conclusions, and by telling the reader about the next idea—what's going to happen, when it has started, and when it is finished. Use these words and phrases to make smooth transitions in your writing:

Alert the reader of a change with words such as *but, yet, however, nevertheless, still,* and *thus.* These words alert readers that their expectations are not likely to be met, thereby preparing them for new information.

Add information with words such as *furthermore, also, similarly,* and *moreover.*

Compare with words such as *similarly, like,* and *likewise.*

Change directions, contrast ideas, and **point out exceptions** with words and phrases such as *but, however, on the contrary, even though, nonetheless, conversely, though, nevertheless, on the other hand, although, alas,* and *yet.*

Caution readers with words such as *however* and *but.*

Illustrate with phrases such as *for example, namely,* and *for instance.*

Qualify a statement with *yet* and *still.*

Concede with words and phrases such as *although, even though, since,* and *though.*

Subordinate ideas with words such as *because, although, while, where,* and *if.*

Indicate time sequences with words such as *first, next, later, finally, when, as, then,* and *while.*

Indicate location with phrases such as *in the next room, down the hall,* and *in the zoology lab.*

Link equal ideas with words such as *and, either/or,* and *neither/nor.*

Indicate cause and effect relationships with words such as *therefore, so, because, thus, hence*, and *consequently.*

Conclude with words such as *therefore, thus, accordingly, moreover*, and *consequently.*

Transitions create dialogue and discussion while keeping the reader oriented to what's going on. Thus, they usually occur at the beginning of sentences and near the middle of paragraphs where examples, comparisons, and cautions are most likely to occur. Without connectors, writing usually crumbles into staccato-like "Me Tarzan. You Jane" prose that annoys, bores, and eventually loses the reader. However, be sure to use connectors correctly: Don't write "but" unless you actually show a contrast or change direction. Signaling a relationship that doesn't exist confuses readers.

Just as few things can improve a sentence more than a well-chosen verb, so too can few things improve a paragraph's cohesiveness more than logic and effective transitions. However, other things also affect cohesion when you write about science. These include trying not to hedge, writing fairly, punctuating your writing correctly, and knowing when to stop writing.

Try Not to Hedge

It is a strange model and embodies several unusual features. However, since DNA is an unusual substance, we are not hesitant in being bold.
—James Watson, writing to a friend one month before the public announcement of the structure of DNA—a discovery many scientists now regard as the most significant since Mendel's

Put a grain of boldness into everything you do.
—Baltasar Gracián

Careful biologists always have some degree of doubt about their conclusions. To express this doubt, biology has its own idiom of caution and confidence. None of us wants to sound like either a smug dogmatist or an uncertain milquetoast. How successfully we walk the line between seeming timidity and arrogance depends in part on how well we hedge with words such as *almost, possibly*, and *perhaps*. Hedge words give us room to backpedal and to make exceptions. However, they can also make us appear wishy-washy. So what's a biologist to do?

Don't mumble. Once you've decided what you want to say, come right out and say it. State points as emphatically as you can support them—no more, no less. Never hesitate to suspend judgment (and do more experiments) when your evidence is inconclusive; such a measured agnosticism is a valuable reflex for a

The Hidden Meanings of Some Hedge Words

It's always wise to state your conclusions as emphatically as you can support them. When you're unsure, don't hedge; just state your level of uncertainty and the reasons behind it as straightforwardly as possible. However, hedging has its place, especially when you need to cover problems in effort or logic. With many people, many hedge words and phrases have taken on a new meaning. For example, consider this list compiled by an unknown author as a guide for interpreting reports:

It has long been known that . . .
> *I haven't bothered to look up the original reference.*

Of great theoretical and practical importance . . .
> *It's interesting to me.*

Nine of the samples were chosen for detailed study . . .
> *The results from other samples didn't make sense, so I ignored them.*

Typical results are shown . . .
> *I've shown the best results.*

These results will be presented at a later date . . .
> *I might get around to this sometime.*

An exhaustive review of the literature shows that . . .
> *I found a 1986 paper that says . . .*

It is believed that . . .
> *I think that . . .*

It is generally believed that . . .
> *A couple of people think that . . .*

It is clear that much additional work is required before a complete understanding of this work will be achieved.
> *I don't understand it.*

Thanks are due to Randy Wayne for assistance with the experiment and to Jerry Hubschman for valuable discussions.
> *Wayne did the work and Hubschman told me what it meant.*

scientist. However, remember that you can't expect others to make up their minds about your work if you haven't made up *your* mind about your work. Therefore, try to come to a firm conclusion.

Use hedge words and phrases such as *often, sometimes, virtually, apparently, seemingly, in some ways, to a certain extent, may, might, perhaps, suggest, is likely to, could, may be possible,* and *probably* only if they're essential. Hedge words such as these are often referred to as "weasel words" because they're used by people trying to weasel their way out of taking a stand. If they're overused, readers feel that the writer is unsure of everything in the paper and refuses to be held accountable. At worst, overusing weasel words makes you appear deceitful, and at best they make you appear wishy-washy. For example, consider this sentence:

> It seems that it could possibly be wise to follow this procedure if a better one is not proposed very soon.

This student uses a quadruple hedge (*seems, could, possibly, if*) before ironically trying to intensify his idea by adding *very*.

Every qualifier whittles away at the reader's trust in you. Don't diminish this trust with a flood of hedge words. Be bold, not *kind of* bold. Do not hesitate to emphatically state conclusions supported by your results. Also remember that many phrases used for emphasis can project arrogance. For example, phrases such as, "As everyone knows," "It is true that," "It is clear that," "As any fool can see," "Of course," "This clearly demonstrates that," "As you should know," and "It is obvious that," rather than ward off challengers, often insult readers. At best, these phrases usually mean little more than "believe me," and they become background static that robs your writing of precision and clarity. At worst, they make you seem arrogant and defensive.

Accepting evasiveness, like accepting pretentiousness, pollutes writing. Don't use a squid's approach to writing and hide behind a sea of ink. Similarly, don't shirk your scientific responsibility of making a judgment. Give readers a take-home message that's logical, convincing, and based on evidence. If your evidence is sufficient to delete qualifiers such as "may," then delete them. If not, ~~it is quite possible that~~ you *may* need to do more research before publishing your work.

Write Fairly

For I am well aware that scarcely a single point is discussed in this volume on which facts cannot be addressed, often apparently leading to conclusions opposite to those at which I have arrived. A fair result can be obtained only by fully stating and balancing the facts and arguments on both sides of each question. . . .
  —Charles Darwin, *The Origin of Species*

He who knows only his own side of the case, knows little of that.
  —John Stuart Mill

Write fairly by acknowledging sources of error, assumptions inherent in your argument, and other possible interpretations of your results. Demonstrate fairness and objectivity not by using passive voice, but by basing your arguments on evidence and logic. Consider all options and approaches and be sure that nothing is implied or left to the readers' imagination. Cite relevant papers even if they don't support your argument. If your argument is sound and you've included evidence to *show* the reader that it's sound, there's no danger that the reader will turn on you. The only danger is that readers will discover the conflict themselves and wonder why it never occurred to you.

Do not assign human traits to other organisms or things. For example, do not write that results suggest or that data indicate. These are human activities that results and data cannot do. Similarly, do not write teleological sentences such as, "Giraffes evolved long necks because they wanted to reach leaves high in trees" or "Birds have wings for flying and a beak for feeding." These sentences state a result as if it were an explanation, goal, cause, or purpose.

Remove biases toward race, disabilities, and sex from your writing.[5] For example, whether used as a prefix (*manpower*) or as a suffix (*chairman*), *man* evokes a strong reaction in many readers. Since *man* can refer both to a male and to both sexes, it produces ambiguity that many readers find unsettling or even offensive. Revise sentences that contain potentially offensive words. You lose nothing and gain much when you do this. For example:

BIASED: Select a spokesman to present the results of your experiment.

NEUTRAL: Select a spokesperson to present the results of your experiment.

NEUTRAL: Select someone to present the results of your group.

Know When to Stop Writing

Not knowing how or when to stop writing is as much a liability in writing as it is in speaking. Consider the simple advice of the King to the White Rabbit:

> . . . go on till you come to the end; then stop. —Lewis Carroll

[5]Many women become brilliant scientists despite the societal and economic barriers to their success. For example, Gerty Cori won a Nobel Prize in 1947 for her studies of sugar metabolism in humans, Dorothy Hodgkin won a Nobel Prize in 1964 for her work with vitamin B_{12} and penicillin, Barbara McClintock in 1983 for discovering mobile elements in genetics, and Rita Levi-Montaleini in 1986 for discovering nerve-growth factor. Many of these discoveries have been based on work of other women. For example, Marie Sklodowska Curie discovered two elements (radium and polonium) and opened the Atomic Age by showing that radiation is an atomic property. In 1903 she became the first woman to win a Nobel Prize and later, in 1911, the first person to win two Nobel Prizes. Her daughter, Irene Joliot-Curie, won a Nobel Prize in 1935 for discovering artificial isotopes. Twenty-two years later, Rosalyn Yallow won a Nobel Prize for developing the radioimmunoassay, a technique sensitive enough to measure a teaspoonful of insulin in a lake 62 miles long, 62 miles wide, and 30 feet deep. Yallow's work was based on artificial isotopes made possible by Irene Curie. Unfortunately, the great work of many women scientists has also been overlooked. For example, Watson's and Crick's Nobel Prize in 1962 was based on analyses and interpretations of data by Rosalind Franklin. Similarly, Candace Pert discovered endorphins, but the Lasker Prize for their discovery went to Solomon Snyder, the director of her lab.

Heed the King's advice: Stop when you've made all of your points. Repetition won't strengthen your argument. When you've said everything highlighted in your outline, and said it well, stop writing. However, remember the words of E.B. White: "When you say something, make sure you have said it. The chances of your having said it are only fair."

There are many ways to end a paper. Look at how Stephen Jay Gould closed an essay on the probability of finding extraterrestrial life:

> Ultimately, however, I must justify the attempt at such a long shot simply by stating that a positive result would be the most cataclysmic event in our entire intellectual history. Curiosity impels, and makes us human. Might it impel others as well?

Wallace Stegner hammers home his message in *The Gift of Wilderness* with this long and powerful sentence:

> Instead of easing air-pollution controls in order to postpone the education of the automobile industry; instead of opening our forests to greatly increased timber cutting; instead of running our national parks to please and profit the concessionaires; instead of violating our wilderness areas by allowing oil and mineral exploration with rigs and roads and seismic detonations, we might bear in mind what those precious places are: playgrounds, schoolrooms, laboratories, yes, but above all shrines, in which we can learn to know both the natural world and ourselves, and be at least half reconciled to what we see.

Compare this with what a confused student wrote when he didn't know quite when to stop writing about heredity:

> Heredity means that if your grandfather didn't have any children, then your father probably wouldn't have any, and neither would you, probably.

If you knew where you were going with your paper, you'll know when you arrive and when you should stop writing. You'll be able to answer yes to all of these questions:

Have you said what you meant to say?

Have you told your readers everything they need to know?

Have you answered an important biological question?

Are the conclusions that you want readers to draw from your evidence clear?

Have you written your paper precisely, clearly, and concisely?

Have you arranged your arguments logically and supported your arguments with evidence?

Have you presented information in digestible chunks?

When you've answered these questions, stop writing.

Punctuate Your Writing Correctly

The workmanlike sentence almost punctuates itself.
   —Wilson Follett

I spent the whole morning putting in a comma; I
spent the whole afternoon taking it out again.
   —Oscar Wilde

No matter what any of the grammar books or
English teachers say, punctuation is an arbitrary
matter. It should be used to make sentences clear.
   —Andy Rooney

SALLY: Show me where you sprinkle in the little
curvy marks.
CHARLIE BROWN: Commas.
SALLY: Whatever.
   —Charles Schultz

Punctuation is for the reader, not the writer: It should clarify the meaning of your writing. Thus, punctuation is to reading as traffic signs and signals are to driving: It tells readers what's coming, when to stop, and when to go. It also tell readers how sentences relate to each other.

Common sense will help you punctuate your writing, but it is no substitute for understanding a few principles of punctuation. Choosing correct punctuation requires as much discretion as choosing words. Although there are about 30 different punctuation marks, you need to know about only a few to write effectively.

Periods (.)

The primary misuse of periods is that most writers don't reach them soon enough. Remember that short sentences dominate effective writing. Use a period after all declarative sentences and abbreviations. For example:

I am not in the least afraid to die.—last words of Charles Darwin

I don't want it.—last words of Marie Curie, when offered a pain-killing injection

I cannot.—last words of Louis Pasteur, when offered a glass of milk

Don't overdo it.—message given by H.L. Mencken to an obituary writer before Mencken's death

Colon (:)

A colon is a mark of anticipation. It has a special function: that of delivering what was invoiced in the preceding words. A colon introduces a clause that explains an idea in the first part of the sentence and tips off the reader about what's next, such as a list, a quotation, or a move from the general to the specific. For example,

> We studied three kinds of marsupials: wombats, bandicoots, and kangaroos.

My brain: It's my second favorite organ. — Woody Allen

That is right: I have now done. — last words of English chemist and biologist Joseph Priestley, after finishing a few corrections in a manuscript

The difference between intelligence and education is this: Intelligence will make you a good living. — Charles F. Kettering

Semicolon (;)

A semicolon is sort of a half-period: It indicates a pause that is stronger than the pause indicated by a comma but weaker than the pause indicated by a period. It indicates that you're weighing two sides of the same problem. A semicolon behaves like a supercomma, separating two or more related clauses and those with internal punctuation where another comma would be confusing.

The semicolon is a convenient device for grouping related elements for which the period is too strong and the comma too weak. For example,

> Cnidarians have no brain, and cnidarian behavior seems to be completely rigid; no one has yet trained a jellyfish.

> There is no known cure for AIDS; it kills thousands of Americans every year.

Anatomy is to physiology as geography to history; it describes the theatre of events. — Jean Fernel

No amount of experimentation can ever prove me right; a single experiment can prove me wrong. — Albert Einstein

The semicolon is one of the least-appreciated punctuation marks. It suggests a close relationship between two independent clauses; it signals contrast without the stopping power of a period; it helps to amplify a point; it nudges the reader onward with a pleasant feeling of anticipation; it tells readers that there is more to come.

Hyphens (-)

Use hyphens in compound adjectives:

> 100-foot-tall tree

> 20-day-old mouse

Also use hyphens to indicate single-bonds, linked amino-acids, and linked nucleotides in polynucleotides:

CH_3-CH_2-OH

Leucine-Proline-Serine-Alanine

pG-A-C-C-T-T-G-C-Gp

Dashes (—)

Use dashes to interrupt thought, to amplify an idea, or to insert ideas for clarity, emphasis, or explanation. For example,

Facts are not science—as the dictionary is not literature.—Martin H. Fischer

Familiarity breeds contempt—and children.—Mark Twain

Quotation Marks (")

Use quotation marks to indicate exact words, to enclose titles, or to indicate a word that is special or being defined. Closing quotation marks go outside of periods and commas:

"God does not cast the die."
Albert Einstein

Parentheses ( )

Use parentheses to insert explanatory or supplemental information or a reference:

We purified the antigens according to methods described by Arlian (1990).

The turquoise streaks in blue cheese and Roquefort are mycelia of certain species of *Penicillium*.

Comma (,)

Commas are the most frequently used and abused punctuation. Many writers avoid commas, while others paint their writing with them. Commas mark a short pause among items that might be confusing if run together. If you use a comma to signal a short pause, you'll usually be correct.

When you begin a sentence with a modifying clause (as I just did), put a comma after the clause. These clauses usually start with words such as *when, as, because, if, since,* and *although.* For example,

Because they shoot a jet of water through the excurrent siphon when molested, tunicates are also called sea squirts.

As crude a weapon as the cave man's club, the chemical barrage has been hurled against the fabric of life.—Rachel Carson, *Silent Spring*

Commas are also used to insert a word or phrase that interrupts a sentence:

Earth, the only truly closed ecosystem any of us knows, is an organism.—Lewis Thomas

Man is developed from an ovule, about 1/25th of an inch in diameter, which differs in no respect from the ovules of other animals. — Charles Darwin

to separate consecutive adjectives:

". . . small, green, and flat leaves . . ."

to separate introductory phrases:

In the main, opera in English is just about as sensible as baseball in Italian. — H.L. Mencken

to separate independent clauses linked by conjunctions such as *and, but, or*, and *for*:

I am about to, or, I am going to die. Either expression is used. — last words of Dominique Bouhors, a French grammarian

A doctor can bury his mistakes, but an architect can only advise his client to plant vines. — Frank Lloyd Wright

I have read all of Darwin's papers, but this one interests me most.

Medical scientists are nice people, but you should not let them treat you. — August Bier

Opportunities are usually disguised as hard work, so most people don't recognize them. — Ann Landers

Science is the father of knowledge, but opinion breeds ignorance. — Hippocrates

We can lick gravity, but sometimes the paperwork is overwhelming. — Wernher Von Braun

and to address someone directly:

Nurse, it was I who discovered that leeches have red blood. — Baron George Cuvier, on his deathbed when the nurse came to apply leeches

Winston, if you were my husband, I should flavor your coffee with poison. — Lady Astor

Madam, if I were your husband, I should drink it. — Winston Churchill

The comma is the most common and versatile punctuation mark, and can greatly affect meaning.[6] You'll best grasp its many uses by noting how good writers use it.

Exclamation Marks (!)

Exclamation marks indicate surprise, disbelief, or other strong emotions.

---

[6]In the 1984 presidential election, Republicans argued for two days about whether this statement in their platform should contain a comma: "The Republicans oppose any attempt to raise taxes which would harm the recovery." Without a comma after *taxes*, they left the door open to increase taxes that they felt would not harm the recovery. However, adding a comma after *taxes* meant that they opposed any taxes because any taxes would harm the recovery.

But I have tol So little donel So much to dol—last words of Alexander Graham Bell, when asked not to hurry his dictation

Like a loud voice, exclamation marks annoy readers when the marks are overused. Most good writers consider exclamation marks a poor substitute for a well-chosen word. Use exclamation marks sparingly, if at all, in your writing. Use truth, logic, and words for emphasis and understatement.

Question Marks (?)

A question mark follows a direct question and replaces one's voice being raised at the end of a question:

What can be more important than the science of life to any intelligent being who has the good fortune to be alive?—Isaac Asimov

Of all these questions the one he asks most insistently is about man. How does he walk? How does the heart pump blood? What happens when he yawns and sneezes? How does a child live in the womb? How does he die of old age? Leonardo discovered a centenarian in a hospital in Florence and waited gleefully for his demise so that he could examine his veins.—Sir Kenneth Clark, referring to Leonardo da Vinci

Simple, well-written sentences are easy to punctuate. However, others are not. For example, try to figure this one out:

That which is is that which is not is not is not that it it is

That sentence is from *Flowers for Algernon* and is punctuated like this:

That which is, is. That which is not, is not. Is not that it? It isl

If you have a question about punctuation, consult one of the books listed in Appendix 1. Also remember that although punctuation is important, it alone can't save a poorly written paper. Don't use punctuation as a Band-Aid when basic surgery is needed. If you're struggling with punctuation, you're probably fighting a losing battle. Since, as Aristophanes said, "You cannot teach a crab to walk straight," scrap the sentence and start over.

Have Someone Else Read What You've Written

It is impossible to write one's best if nobody else
ever has a look at it.
   —C.S. Lewis

Find the grain of truth in criticism—chew it and
swallow it.
   —D. Sutten

To avoid criticism, do nothing, say nothing, be nothing.
    —Elbert Hubbard

Few things will improve your writing as much as constructive criticism. Therefore, always have someone read what you've written. Ask for criticisms and be willing to consider what the reader says. Although the criticisms may sometimes sting, they'll help you produce a better paper.

Garrett Hardin

The Tragedy of the Commons

Early in his career, Garrett Hardin (b. 1915) studied how to grow algae as a large-scale source of food. However, Hardin then realized that increasing the food supply would increase the size of the population, thus increasing population-related problems. This realization prompted Hardin to write "The Tragedy of the Commons," a famous and controversial essay exploring the ethical and social implications of biological research.

At the end of a thoughtful article on the future of nuclear war, Wiesner and York[1] concluded that: "Both sides in the arms race are . . . confronted by the dilemma of steadily increasing military power and steadily decreasing national security. *It is our considered professional judgment that this dilemma has no technical solution.* If the great powers continue to look for solutions in the area of science and technology only, the result will be to worsen the situation."

I would like to focus your attention not on the subject of the article (national security in a nuclear world) but on the kind of conclusion they reached, namely that there is no technical solution to the problem. An implicit and almost universal assumption of discussions published in professional and semipopular scientific journals is that the problem under discussion has a technical solution. A technical solution may be defined as one that requires a change only in the techniques of the natural sciences, demanding little or nothing in the way of change in human values or ideas of morality.

In our day (though not in earlier times) technical solutions are always welcome. Because of previous failures in prophecy, it takes courage to assert that a desired technical solution is not possible. Wiesner and York exhibited this courage; publishing in a science journal, they insisted that the solution to the problem was not to be found in the natural sciences. They cautiously qualified their statement with the phrase, "It is our considered professional judgment. . . ." Whether they were right or not

---

[1]J.B. Wiesner and H.F. York, *Sci. Amer.* 211 (No. 4), 27 (1964).

is not the concern of the present article. Rather, the concern here is with the important concept of a class of human problems which can be called "no technical solution problems," and, more specifically, with the identification and discussion of one of these.

It is easy to show that the class is not a null class. Recall the game of tick-tack-toe. Consider the problem, "How can I win the game of tick-tack-toe?" It is well known that I cannot, if I assume (in keeping with the conventions of game theory) that my opponent understands the game perfectly. Put another way, there is no "technical solution" to the problem. I can win only by giving a radical meaning to the word "win." I can hit my opponent over the head; or I can drug him; or I can falsify the records. Every way in which I "win" involves, in some sense, an abandonment of the game, as we intuitively understand it. (I can also, of course, openly abandon the game—refuse to play it. This is what most adults do.)

The class of "No technical solution problems" has members. My thesis is that the "population problem," as conventionally conceived, is a member of this class. How it is conventionally conceived needs some comment. It is fair to say that most people who anguish over the population problem are trying to find a way to avoid the evils of over-population without relinquishing any of the privileges they now enjoy. They think that farming the seas or developing new strains of wheat will solve the problem—technologically. I try to show here that the solution they seek cannot be found. The population problem cannot be solved in a technical way, any more than can the problem of winning the game of tick-tack-toe.

Population, as Malthus said, naturally tends to grow "geometrically," or, as we would now say, exponentially. In a finite world this means that the per capita share of the world's goods must steadily decrease. Is ours a finite world?

A fair defense can be put forward for the view that the world is infinite; or that we do not know that it is not. But, in terms of the practical problems that we must face in the next generations with the foreseeable technology, it is clear that we will greatly increase human misery if we do not, during the immediate future, assume that the world available to the terrestrial human population is finite. "Space" is no escape.[2]

A finite world can support only a finite population; therefore, population growth must eventually equal zero. (The case of perpetual wide fluctuations above and below zero is a trivial variant that need not be discussed.) When this condition is met, what will be the situation of mankind? Specifically, can Bentham's goal of "the greatest good for the greatest number" be realized?

No—for two reasons, each sufficient by itself. The first is a

[2]G. Hardin, *J. Hered.* 50, 68 (1959); S. von Hoernor, *Science* 137, 18 (1962).

theoretical one. It is not mathematically possible to maximize for two (or more) variables at the same time. This was clearly stated by von Neumann and Morgenstern,[3] but the principle is implicit in the theory of partial differential equations, dating back at least to D'Alembert (1717–1783).

The second reason springs directly from biological facts. To live, any organism must have a source of energy (for example, food). This energy is utilized for two purposes: mere maintenance and work. For man, maintenance of life requires about 1600 kilocalories a day ("maintenance calories"). Anything that he does over and above merely staying alive will be defined as work, and is supported by "work calories" which he takes in. Work calories are used not only for what we call work in common speech; they are also required for all forms of enjoyment, from swimming and automobile racing to playing music and writing poetry. If our goal is to maximize population it is obvious what we must do: We must make the work calories per person approach as close to zero as possible. No gourmet meals, no vacations, no sports, no music, no literature, no art. . . . I think that everyone will grant, without argument or proof, that maximizing population does not maximize goods. Bentham's goal is impossible.

In reaching this conclusion I have made the usual assumption that it is the acquisition of energy that is the problem. The appearance of atomic energy has led some to question this assumption. However, given an infinite source of energy, population growth still produces an inescapable problem. The problem of the acquisition of energy is replaced by the problem of its dissipation, as J.H. Fremlin has so wittily shown.[4] The arithmetic signs in the analysis are, as it were, reversed; but Bentham's goal is still unobtainable.

The optimum population is, then, less than the maximum. The difficulty of defining the optimum is enormous; so far as I know, no one has seriously tackled this problem. Reaching an acceptable and stable solution will surely require more than one generation of hard analytical work—and much persuasion.

We want the maximum good per person; but what is good? To one person it is wilderness, to another it is ski lodges for thousands. To one it is estuaries to nourish ducks for hunters to shoot; to another it is factory land. Comparing one good with another is, we usually say, impossible because goods are incommensurable. Incommensurables cannot be compared.

Theoretically this may be true; but in real life incommensurables *are* commensurable. Only a criterion of judgment and a system of weighting are needed. In nature the criterion is survival. Is it better for a species to be small and hideable, or large and powerful? Natural

---

[3]J. von Neumann and O. Morgenstern, *Theory of Games and Economic Behavior* (Princeton Univ. Press, Princeton, N.J., 1947), p. 11.

[4]J.H. Fremlin, *New Sci.*, No. 415 (1964), p. 285.

selection commensurates the incommensurables. The compromise achieved depends on a natural weighting of the values of the variables.

Man must imitate this process. This is no doubt that in fact he already does, but unconsciously. It is when the hidden decisions are made explicit that the arguments begin. The problem for the years ahead is to work out an acceptable theory of weighting. Synergistic effects, nonlinear variation, and difficulties in discounting the future make the intellectual problem difficult, but not (in principle) insoluble.

Has any cultural group solved this practical problem at the present time, even on an intuitive level? One simple fact proves that none has: there is no prosperous population in the world today that has, and has had for some time, a growth rate of zero. Any people that has intuitively identified its optimum point will soon reach it, after which its growth rate becomes and remains zero.

Of course, a positive growth rate might be taken as evidence that a population is below its optimum. However, by any reasonable standards, the most rapidly growing populations on earth today are (in general) the most miserable. This association (which need not be invariable) casts doubt on the optimistic assumption that the positive growth rate of a population is evidence that it has yet to reach its optimum.

We can make little progress in working toward optimum population size until we explicitly exorcize the spirit of Adam Smith in the field of practical demography. In economic affairs, *The Wealth of Nations* (1776) popularized the "invisible hand," the idea that an individual who "intends only his own gain," is, as it were, "led by an invisible hand to promote . . . the public interest."[5] Adam Smith did not assert that this was invariably true, and perhaps neither did any of his followers. But he contributed to a dominant tendency of thought that has ever since interfered with positive action based on rational analysis, namely, the tendency to assume that decisions reached individually will, in fact, be the best decisions for an entire society. If this assumption is correct it justifies the continuance of our present policy of laissez-faire in reproduction. If it is correct we can assume that men will control their individual fecundity so as to produce the optimum population. If the assumption is not correct, we need to reexamine our individual freedoms to see which ones are defensible.

Tragedy of Freedom in a Commons

The rebuttal to the invisible hand in population control is to be found in a scenario first sketched in a little-known pamphlet[6] in 1833 by a

---

[5] A. Smith, *The Wealth of Nations* (Modern Library, New York, 1937), p. 423.

[6] W.F. Lloyd, *Two Lectures on the Checks to Population* (Oxford Univ. Press, Oxford, England, 1833), reprinted (in part) in *Population, Evolution, and Birth Control*, G. Hardin, Ed. (Freeman, San Francisco, 1964), p. 37.

mathematical amateur named William Forster Lloyd (1794–1852). We may well all it "the tragedy of the commons," using the word "tragedy" as the philosopher Whitehead used it:[7] "The essence of dramatic tragedy is not unhappiness. It resides in the solemnity of the remorseless working of things." He then goes on to say, "This inevitableness of destiny can only be illustrated in terms of human life by incidents which in fact involve unhappiness. For it is only by them that the futility of escape can be made evident in the drama."

The tragedy of the commons develops in this way. Picture a pasture open to all. It is to be expected that each herdsman will try to keep as many cattle as possible on the commons. Such an arrangement may work reasonably satisfactorily for centuries because tribal wars, poaching, and disease keep the numbers of both man and beast well below the carrying capacity of the land. Finally, however, comes the day of reckoning, that is, the day when the long-desired goal of social stability becomes a reality. At this point, the inherent logic of the commons remorselessly generates tragedy.

As a rational being, each herdsman seeks to maximize his gain. Explicitly or implicitly, more or less consciously, he asks, "What is the utility *to me* of adding one more animal to my herd?" This utility has one negative and one positive component.

1. The positive component is a function of the increment of one animal. Since the herdsman receives all the proceeds from the sale of the additional animal, the positive utility is nearly +1.

2. The negative component is a function of the additional overgrazing created by one more animal. Since, however, the effects of overgrazing are shared by all the herdsmen, the negative utility for any particular decision-making herdsman is only a fraction of −1.

Adding together the component partial utilities, the rational herdsman concludes that the only sensible course for him to pursue is to add another animal to his herd. And another; and another. . . . But this is the conclusion reached by each and every rational herdsman sharing a commons. Therein is the tragedy. Each man is locked into a system that compels him to increase his herd without limit—in a world that is limited. Ruin is the destination toward which all men rush, each pursuing his own best interest in a society that believes in the freedom of the commons. Freedom in a commons brings ruin to all.

Some would say that this is a platitude. Would that it were! In a sense, it was learned thousands of years ago, but natural selection favors the forces of psychological denial.[8] The individual benefits as an individual from his ability to deny the truth even though society as a

---

[7]A.N. Whitehead, *Science and the Modern World* (Mentor, New York, 1948), p. 17.
[8]G. Hardin, Ed. *Population, Evolution, and Birth Control* (Freeman, San Francisco, 1964), p. 56.

whole, of which he is a part, suffers. Education can counteract the natural tendency to do the wrong thing, but the inexorable succession of generations requires that the basis for this knowledge be constantly refreshed.

A simple incident that occurred a few years ago in Leominster, Massachusetts, shows how perishable the knowledge is. During the Christmas shopping season the parking meters downtown were covered with plastic bags that bore tags reading: "Do not open until after Christmas. Free parking courtesy of the mayor and city council." In other words, facing the prospect of an increased demand for already scarce space, the city fathers reinstituted the system of the commons. (Cynically, we suspect that they gained more votes than they lost by this retrogressive act.)

In an approximate way, the logic of the commons has been understood for a long time, perhaps since the discovery of agriculture or the intervention of private property in real estate. But it is understood mostly only in special cases which are not sufficiently generalized. Even at this late date, cattlemen leasing national land on the western ranges demonstrate no more than an ambivalent understanding, in constantly pressuring federal authorities to increase the head count to the point where over-grazing produces erosion and weed-dominance. Likewise, the oceans of the world continue to suffer from the survival of the philosophy of the commons. Maritime nations still respond automatically to the shibboleth of the "freedom of the seas." Professing to believe in the "inexhaustible resources of the oceans," they bring species after species of fish and whales closer to extinction.[9]

The National Parks present another instance of the working out of the tragedy of the commons. At present, they are open to all, without limit. The parks themselves are limited in extent—there is only one Yosemite Valley—whereas population seems to grow without limit. The values that visitors seek in the parks are steadily eroded. Plainly, we must soon cease to treat the parks as commons or they will be of no value to anyone.

What shall we do? We have several options. We might sell them off as private property. We might keep them as public property, but allocate the right to enter them. The allocation might be on the basis of wealth, by the use of an auction system. It might be on the basis of merit, as defined by some agreed-upon standards. It might be by lottery. Or it might be on a first-come, first-served basis, administered to long queues. These, I think, are all the reasonable possibilities. They are all objectionable. But we must choose—or acquiesce in the destruction of the commons that we call our National Parks.

[9]S. McVay, *Sci. Amer.* 216 (No. 8), 13 (1966).

Pollution

In a reverse way, the tragedy of the commons reappears in problems of pollution. Here it is not a question of taking something out of the commons, but of putting something in — sewage, or chemical, radioactive, and heat wastes into water; noxious and dangerous fumes into the air; and distracting and unpleasant advertising signs into the line of sight. The calculations of utility are much the same as before. The rational man finds that his share of the cost of the wastes he discharges into the commons is less than the cost of purifying his wastes before releasing them. Since this is true for everyone, we are locked into a system of "fouling our own nest," so long as we behave only as independent, rational, free-enterprisers.

The tragedy of the commons as a food basket is averted by private property, or something formally like it. But the air and waters surrounding us cannot readily be fenced, and so the tragedy of the commons as a cesspool must be prevented by different means, by coercive laws or taxing devices that make it cheaper for the polluter to treat his pollutants than to discharge them untreated. We have not progressed as far with the solution of this problem as we have with the first. Indeed, our particular concept of private property, which deters us from exhausting the positive resources of the earth, favors pollution. The owner of a factory on the bank of a stream — whose property extends to the middle of the stream — often has difficulty seeing why it is not his natural right to muddy the waters flowing past his door. The law, always behind the times, requires elaborate stitching and fitting to adapt it to this newly perceived aspect of the commons.

The pollution problem is a consequence of population. It did not much matter how a lonely American frontiersman disposed of his waste. "Flowing water purifies itself every 10 miles," my grandfather used to say, and the myth was near enough to the truth when he was a boy, for there were not too many people. But as population became denser, the natural chemical and biological recycling processes became overloaded, calling for a redefinition of property rights.

How to Legislate Temperance?

Analysis of the pollution problem as a function of population density uncovers a not generally recognized principle of morality, namely: *the morality of an act is a function of the state of the system at the time it is performed.*[10] Using the commons as a cesspool does not harm the

[10]J. Fletcher, *Situation Ethics* (Westminister, Philadelphia, 1966).

general public under frontier conditions, because there is no public; the same behavior in a metropolis is unbearable. A hundred and fifty years ago a plainsman could kill an American bison, cut out only the tongue for his dinner, and discard the rest of the animal. He was not in any important sense being wasteful. Today, with only a few thousand bison left, we would be appalled at such behavior.

In passing, it is worth noting that the morality of an act cannot be determined from a photograph. One does not know whether a man killing an elephant or setting fire to the grassland is harming others until one knows the total system in which his act appears. "One picture is worth a thousand words," said an ancient Chinese; but it may take 10,000 words to validate it. It is as tempting to ecologists as it is to reformers in general to try to persuade others by way of the photographic shortcut. But the essence of an argument cannot be photographed: it must be presented rationally—in words.

That morality is system-sensitive escaped the attention of most codifiers of ethics in the past. "Thou shalt not . . ." is the form of traditional ethical directives which make no allowance for particular circumstances. The laws of our society follow the pattern of ancient ethics, and therefore are poorly suited to governing a complex, crowded, changeable world. Our epicyclic solution is to augment statutory law with administrative law. Since it is practically impossible to spell out all the conditions under which it is safe to burn trash in the back yard or to run an automobile without smog-control, by law we delegate the details to bureaus. The result is administrative law, which is rightly feared for an ancient reason—*Quis custodiet ipsos custodes?*—"Who shall watch the watchers themselves?" John Adams said that we must have "a government of laws and not men." Bureau administrators, trying to evaluate the morality of acts in the total system, are singularly liable to corruption, producing a government by men, not laws.

Prohibition is easy to legislate (though not necessarily to enforce); but how do we legislate temperance? Experience indicates that it can be accomplished best through the mediation of administrative law. We limit possibilities unnecessarily if we suppose that the sentiment of *Quis custodiet* denies us the use of administrative law. We should rather retain the phrase as a perpetual reminder of fearful dangers we cannot avoid. The great challenge facing us now is to invent the corrective feedbacks that are needed to keep custodians honest. We must find ways to legitimate the needed authority of both the custodians and the corrective feedbacks.

Freedom To Breed Is Intolerable

The tragedy of the commons is involved in population problems in another way. In a world governed solely by the principle of "dog eat dog" — if indeed there ever was such a world — how many children a family had would not be a matter of public concern. Parents who bred too exuberantly would leave fewer descendants, not more, because they would be unable to care adequately for their children. David Lack and others have found that such a negative feedback demonstrably controls of fecundity of birds.[11] But men are not birds, and have not acted like them for millenniums, at least.

If each human family were dependent only on its own resources; *if* the children of improvident parents starved to death; *if,* thus, over-breeding brought its own "punishment" to the germ line — *then* there would be no public interest in controlling the breeding of families. But our society is deeply committed to the welfare state,[12] and hence is confronted with another aspect of the tragedy of the commons.

In a welfare state, how shall we deal with the family, the religion, the race, or the class (or indeed any distinguishable and cohesive group) that adopts overbreeding as a policy to secure its own aggrandizement?[13] To couple the concept of freedom to breed with the belief that everyone born has an equal right to the commons is to lock the world into a tragic course of action.

Unfortunately this is just the course of action that is being pursued by the United Nations. In late 1967, some 30 nations agreed to the following:[14]

> The Universal Declaration of Human Rights describes the family as the natural and fundamental unit of society. It follows that any choice and decision with regard to the size of the family must irrevocably rest with the family itself, and cannot be made by anyone else.

It is painful to have to deny categorically the validity of this right; denying it, one feels as uncomfortable as a resident of Salem, Massachusetts, who denied the reality of witches in the 17th century. At the present time, in liberal quarters, something like a taboo acts to inhibit criticism of the United Nations. There is a feeling that the United Nations is "our last and best hope," that we shouldn't find fault with it; we shouldn't play into the hands of the archconservatives. However, let us not forget what Robert Louis Stevenson said: "The truth that is suppressed by friends is the readiest weapon of the enemy." If we love the truth we must openly deny the validity of the Universal Declaration of Human Rights, even though it is promoted by the United Nations. We should also join with Kingsley Davis[15]

[11]D. Lack, *The Natural Regulation of Animal Numbers* (Clarendon Press, Oxford, 1954).

[12]H. Girvetz, *From Wealth to Welfare* (Stanford Univ. Press, Stanford, Calif., 1950).

[13]G. Hardin, *Perspec. Biol. Med.* 6, 366 (1963).

[14]U. Thant, *Int. Planned Parenthood News*, No. 168 (February 1968), p. 3.

[15]K. Davis, *Science* 158, 730 (1967).

in attempting to get Planned Parenthood-World Population to see the error of its ways in embracing the same tragic ideal.

Conscience Is Self-Eliminating

It is a mistake to think that we can control the breeding of mankind in the long run by an appeal to conscience. Charles Galton Darwin made this point when he spoke on the centennial of the publication of his grandfather's great book. The argument is straightforward and Darwinian.

People vary. Confronted with appeals to limit breeding, some people will undoubtedly respond to the plea more than others. Those who have more children will produce a larger fraction of the next generation than those with more susceptible consciences. The difference will be accentuated, generation by generation.

In C. G. Darwin's words: "It may well be that it would take hundreds of generations for the progenitive instinct to develop in this way, but if it should do so, nature would have taken her revenge, and the variety *Homo contracipiens* would become extinct and would be replaced by the variety *Homo progenitivus.*"[16]

The argument assumes that conscience or the desire for children (no matter which) is hereditary—but hereditary only in the most general formal sense. The result will be the same whether the attitude is transmitted through germ cells, or exosomatically, to use A. J. Lotka's term. (If one denies the latter possibility as well as the former, then what's the point of education?) The argument has here been stated in the context of the population problem, but it applies equally well to any instance in which society appeals to an individual exploiting a commons to restrain himself for the general good—by means of his conscience. To make such an appeal is to set up a selective system that works toward the elimination of conscience from the race.

Pathogenic Effects of Conscience

The long-term disadvantage of an appeal to conscience should be enough to condemn it; but has serious short-term disadvantages as well. If we ask a man who is exploiting a commons to desist "in the name of conscience," what are we saying to him? What does he hear?—not only at the moment but also in the wee small hours of the night when, half asleep, he remembers not merely the words we used but also the

[16]S. Tax, Ed., *Evolution after Darwin* (Univ. of Chicago Press, Chicago, 1960), vol. 2, p. 469.

nonverbal communication cues we gave him unawares? Sooner or later, consciously or subconsciously, he senses that he has received two communications, and that they are contradictory: (i) (intended communication) "If you don't do as we ask, we will openly condemn you for not acting like a responsible citizen"; (ii) (the unintended communication) "If you *do* behave as we ask, we will secretly condemn you for a simpleton who can be shamed into standing aside while the rest of us exploit the commons."

Everyman then is caught in what Bateson has called a "double bind." Bateson and his co-workers have made a plausible case for viewing the double bind as an important causative factor in the genesis of schizophrenia.[17] The double bind may not always be so damaging, but it always endangers the mental health of anyone to whom it is applied. "A bad conscience," said Nietzsche, "is a kind of illness."

To conjure up a conscience in others is tempting to anyone who wishes to extend his control beyond the legal limits. Leaders at the highest level succumb to this temptation. Has any President during the past generation failed to call on labor unions to moderate voluntarily their demands for higher wages, or to steel companies to honor voluntary guidelines on prices? I can recall none. The rhetoric used on such occasions is designed to produce feelings of guilt in noncooperators.

For centuries it was assumed without proof that guilt was a valuable, perhaps even an indispensable, ingredient of the civilized life. Now, in this post-Freudian world, we doubt it.

Paul Goodman speaks from the modern point of view when he says: "No good has ever come from feeling guilty, neither intelligence, policy, nor compassion. The guilty do not pay attention to the object but only to themselves, and not even to their own interests, which might make sense, but to their anxieties."[18]

One does not have to be a professional psychiatrist to see the consequences of anxiety. We in the Western world are just emerging from a dreadful two-centuries-long Dark Ages of Eros that was sustained partly by prohibition laws, but perhaps more effectively by the anxiety-generating mechanisms of education. Alex Comfort has told the story well in *The Anxiety Makers*;[19] it is not a pretty one.

Since proof is difficult, we may even concede that the results of anxiety may sometimes, from certain points of view, be desirable. The larger question we should ask is whether, as a matter of policy, we should ever encourage the use of a technique the tendency (if not the

---

[17]G. Bateson, D. D. Jackson, J. Haley, J. Weakland, *Behav. Sci.* 1, 251 (1956).
[18]P. Goodman, *New York Rev. Books* 10(8), 22 (23 May 1968).
[19]A. Comfort, *The Anxiety Makers* (Nelson, London, 1967).

intention) of which is psychologically pathogenic. We hear much talk these days of responsible parenthood; the coupled words are incorporated into the titles of some organizations devoted to birth control. Some people have proposed massive propaganda campaigns to instill responsibility into the nation's (or the world's) breeders. But what is the meaning of the word responsibility in this context? Is it not merely a synonym for the word conscience? When we use the word responsibility in the absence of substantial sanctions are we not trying to browbeat a free man in a commons into acting against his own interest? Responsibility is a verbal counterfeit for a substantial *quid pro quo*. It is an attempt to get something for nothing.

If the word responsibility is to be used at all, I suggest that it be in the sense Charles Frankel uses it.[20] "Responsibility," says this philosopher, "is the product of definite social arrangements." Notice that Frankel calls for social arrangements—not propaganda.

Mutual Coercion Mutually Agreed Upon

The social arrangements that produce responsibility are arrangements that create coercion, of some sort. Consider bank-robbing. The man who takes money from a bank acts as if the bank were a commons. How do we prevent such action? Certainly not by trying to control his behavior solely by a verbal appeal to his sense of responsibility. Rather than rely on propaganda we follow Frankel's lead and insist that a bank is not a commons; we seek the definite social arrangements that will keep it from becoming a commons. That we thereby infringe on the freedom of would-be robbers we neither deny nor regret.

The morality of bank-robbing is particularly easy to understand because we accept complete prohibition of this activity. We are willing to say "Thou shalt not rob banks," without providing for exceptions. But temperance also can be created by coercion. Taxing is a good coercive device. To keep downtown shoppers temperate in their use of parking space we introduce parking meters for short periods, and traffic fines for longer ones. We need not actually forbid a citizen to park as long as he wants to; we need merely make it increasingly expensive for him to do so. Not prohibition, but carefully biased options are what we offer him. A Madison Avenue man might call this persuasion; I prefer the greater candor of the word coercion.

Coercion is a dirty word to most liberals now, but it need not forever be so. As with the four-letter words, its dirtiness can be cleansed away by exposure to the light, by saying it over and over without apology or embarrassment. To many, the word coercion implies arbitrary

---

[20]C. Frankel, *The Case for Modern Man* (Harper, New York, 1955), p. 203.

decisions of distant and irresponsible bureaucrats; but this is not a necessary part of its meaning. The only kind of coercion I recommend is mutual coercion, mutually agreed upon by the majority of the people affected.

To say that we mutually agree to coercion is not to say that we are required to enjoy it, or even to pretend to enjoy it. Who enjoys taxes? We all grumble about them. But we accept compulsory taxes because we recognize that voluntary taxes would favor the conscienceless. We institute and (grumblingly) support taxes and other coercive devices to escape the horror of the commons.

An alternative to the commons need not be perfectly just to be preferable. With real estate and other material goods, the alternative we have chosen is the institution of private property coupled with legal inheritance. Is this system perfectly just? As a genetically trained biologist I deny that it is. It seems to me that, if there are to be differences in individual inheritance, legal possession should be perfectly correlated with biological inheritance—that those who are biologically more fit to be the custodians of property and power should legally inherit more. But genetic recombination continually makes a mockery of the doctrine of "like father, like son" implicit in our laws of legal inheritance. An idiot can inherit millions, and a trust fund can keep his estate intact. We must admit that our legal system of private property plus inheritance is unjust—but we put up with it because we are not convinced, at the moment, that anyone has invented a better system. The alternative of the commons is too horrifying to contemplate. Injustice is preferable to total ruin.

It is one of the peculiarities of the warfare between reform and the status quo that it is thoughtlessly governed by a double standard. Whenever a reform measure is proposed it is often defeated when its opponents triumphantly discover a flaw in it. As Kingsley Davis has pointed out,[21] worshippers of the status quo sometimes imply that no reform is possible without unanimous agreement, an implication contrary to historical fact. As nearly as I can make out, automatic rejection of proposed reforms is based on one of two unconscious assumptions: (i) that the status quo is perfect; or (ii) that the choice we face is between reform and no action; if the proposed reform is imperfect, we presumably should take no action at all, while we wait for a perfect proposal.

But we can never do nothing. That which we have done for thousands of years is also action. It also produces evils. Once we are aware that the status quo is action, we can then compare its discoverable advantages and disadvantages with the predicted advantages and disadvantages of the proposed reform, discounting as best we can for

[21]J. D. Roslansky, *Genetics and the Future of Man* (Appleton-Century-Crofts, New York, 1966), p. 177.

our lack of experience. On the basis of such a comparison, we can make a rational decision which will not involve the unworkable assumption that only perfect systems are tolerable.

Recognition of Necessity

Perhaps the simplest summary of this analysis of man's population problems is this: the commons, if justifiable at all, is justifiable only under conditions of low-population density. As the human population has increased, the commons has had to be abandoned in one aspect after another.

First we abandoned the commons in food gathering, enclosing farm land and restricting pastures and hunting and fishing areas. These restrictions are still not complete throughout the world.

Somewhat later we saw that the commons as a place for waste disposal would also have to be abandoned. Restrictions on the disposal of domestic sewage are widely accepted in the Western world; we are still struggling to close the commons to pollution by automobiles, factories, insecticide sprayers, fertilizing operations, and atomic energy installations.

In a still more embryonic state is our recognition of the evils of the commons in matters of pleasure. There is almost no restriction on the propagation of sound waves in the public medium. The shopping public is assaulted with mindless music, without its consent. Our government is paying out billions of dollars to create supersonic transport which will disturb 50,000 people for every one person who is whisked from coast to coast 3 hours faster. Advertisers muddy the airwaves of radio and television and pollute the view of travelers. We are a long way from outlawing the commons in matters of pleasure. Is this because our Puritan inheritance makes us view pleasure as something of a sin, and pain (that is, the pollution of advertising) as the sign of virtue?

Every new enclosure of the commons involves the infringement of somebody's personal liberty. Infringements made in the distant past are accepted because no contemporary complains of a loss. It is the newly proposed infringements that we vigorously oppose; cries of "rights" and "freedom" fill the air. But what does "freedom" mean? When men mutually agreed to pass laws against robbing, mankind became more free, not less so. Individuals locked into the logic of the commons are free only to bring on universal ruin; once they see the necessity of mutual coercion, they become free to pursue other goals. I believe it was Hegel who said, "Freedom is the recognition of necessity."

The most important aspect of necessity that we must now recognize, is the necessity of abandoning the commons in breeding. No technical solution can rescue us from the misery of overpopulation. Freedom to

Maria Rillo

breed will bring ruin to all. At the moment, to avoid hard decisions many of us are tempted to propagandize for conscience and responsible parenthood. The temptation must be resisted, because an appeal to independently acting consciences selects for the disappearance of all conscience in the long run, and an increase in anxiety in the short.

The only way we can preserve and nurture other and more precious freedoms is by relinquishing the freedom to breed, and that very soon. "Freedom is the recognition of necessity"—and it is the role of education to reveal to all the necessity of abandoning the freedom to breed. Only so, can we put an end to this aspect of the tragedy of the commons.

Why do you think Hardin introduced topics in the order that he did? Write a one-page essay that argues Hardin's position. Then write a one-page essay disputing Hardin's position. Which was easiest for you to write? Why?

Exercises

1. Rachel Carson's *The Sea Around Us* won a National Book Award and went into 11 printings the year it was published (1951). Here is one of its paragraphs:

    > Between the sunlit surface waters of the open sea and the hidden hills and valleys of the ocean floor lies the least known region of the sea. These deep, dark waters, with all their mysteries and their unsolved problems, cover a very considerable part of the earth. The whole world ocean extends over about three-fourths of the surface of the globe. If we subtract the shallow areas of the continental shelves and the scattered banks and shoals, where at least the pale ghost of sunlight moves over the underlying bottom, there still remains about half the earth that is covered by miles-deep, lightless water, that has been dark since the world began.[7]

    What gives this paragraph its cohesion and rhythm?

2. Eliminating unnecessary words and phrases often produces a series of short, choppy sentences. To make your paper readable, you must merge these sentences into cohesive sentences and paragraphs. Combine each of the following sets of sentences into a well-written sentence or paragraph:

    All of the glassware should be cleaned and rinsed.

    The glassware is not disposable.

    The cleaning and rinsing should be thorough.

[7]Carson, Rachel L. 1951. *The Sea Around Us.* New York: Oxford University Press, p. 37.

Wash the glassware with nitric acid.

The acid should be a solution of 50%.

Model: All nondisposable glassware should be thoroughly cleaned and rinsed with a 50% solution of nitric acid.

The biologist is studying the effects of alcohol on the circulatory systems of rats.
The biologist is Dr. Dennis Clark.

The project will study DNA in chimpanzees.
The chimpanzees are native to Africa.
DNA is the genetic building-block.

Antibodies protect us.
Antibodies protect us against disease.
Antibodies attack disease antigens.

The 1991 guidelines of the EPA are available.
The guidelines are for air quality.
The levels for air quality are tolerable levels.
The levels are for each type of pollutant in the air.
The levels vary with toxicity of each pollutant.

3. Improve the readability and clarity of these sentences and paragraphs:

Enclosed please find material which is descriptive of the graduate programs in the Department of Zoology and Wildlife Science at Auburn University. I would deeply appreciate your posting of this material and calling it to the attention of any students who might have an interest in our program.

We suggest it is quite possible that the solutions may have been contaminated.

The fact that bacteria are dispersed nonrandomly seems to possibly suggest that they may be able to perceive magnetic fields.

The results would seem to suggest the possible presence of a new life form.

It is possible that death may occur.

The data suggest that the rats may be abnormal.

The bacterium has about 100 to 200 flagella.

In the majority of instances, the data provided by direct and causal examination of living material under the lens of the optical microscope are inadequate and insufficient for the proper and correct identification of bacteria.

Our proposal follows the sequential itemization of points occurring elsewhere in your RFP, wherever possible, to facilitate your review.

More than mere numbers are required for this experiment.

It is most useful to keep constantly in mind that the term diabetes mellitus refers to a broad and expansive array of health disorders.

There has been an affirmative decision for program termination.

An evaluation of the experiment by us will assure greater efficiency in utilization of experimental animals.

Pursuant to the recent memorandum issued 21 June 1991, because of financial exigencies, it is incumbent upon us all to endeavor to make maximal utilization of telephonic communication in lieu of personal visitation.

Imagine a picture of someone engaged in the activity of trying to learn how to operate a pH meter.

During that period of time, the larynx became blue in color and shiny in appearance.

The instructions must be followed in an accurate manner.

As per our aforementioned discussion, I am herewith enclosing a report of our results.

In my personal opinion, we should basically listen to and think over in a punctilious manner each and every suggestion that is offered to us.

4. Discuss, support, or refute the ideas expressed in these quotations:

Freedom to breed will bring ruin to all. — Garrett Hardin

Few scientists are willing to jeopardize their research funds by publicly criticizing EPA's interpretation of the scientific record. — Roy Gould

Ontogeny recapitulates phylogeny. — Ernest Heinrich Haeckel

The significant chemicals of living tissue are rickety and unstable, which is exactly what is needed for life. — Isaac Asimov

The main conclusion arrived at in this work, namely, that man is descended from some lowly organized form, will, I regret to think, be highly distasteful to many. But there can hardly be a doubt that we are descended from barbarians. — Charles Darwin

I have long felt that biology ought to seem as exciting as a mystery story, for a mystery story is exactly what biology is. — Richard Dawkins, *The Selfish Gene*

5. Read a biology-related article from a recent issue of *Scientific American*. Reduce each paragraph to one sentence.

UNIT TWO

WRITING FOR
YOUR
AUDIENCE

Anyone who writes about science must know
about science, which cuts down competition
considerably.
—Isaac Asimov

First, have something to say; second, say it; third,
stop when you have said it; and finally, give it an
accurate title.
—John Shaw Billings

We have multitudes of facts, but we require, as
they accumulate, organization of them into higher
knowledge; we require generalizations and working
hypotheses.
—Hughlings Jackson

A word to the wise is not sufficient if it doesn't
make any sense.
—James Thurber

Writing comes more easily if you have something
to say.
    —Sholem Asch

The difficulty is not to write but to write what you
mean, not to affect your reader but to affect him
precisely as you wish.
    —Robert Louis Stevenson

If any man wishes to have a clear style, let him first
be clear in his thoughts.
    —Johann Wolfgang von Goethe

Scientific research begins with a set of sentences
which point the way to certain observations and
experiments, the results of which do not become
fully scientific until they have been turned back into
language, yielding again a set of sentences which
then become the basis for further exploration into
the unknown. . . .
    —Benjamin Lee Whorf

I have learned that when I write a research paper I
do far more than summarize conclusions already
neatly stored in my mind. Rather, the writing
process is where I carry out the final comprehension, analysis, and synthesis of my results.
    —Sidney Perkowitz

CHAPTER SIX

Writing for a

General Audience

I believe — as Galileo did when he wrote his two greatest works as dialogues in Italian rather than didactic treatises in Latin, as Thomas Henry Huxley did when he composed his masterful prose free of jargon, as Darwin did when he published all his books for general audiences — that we can still have a genre of science books suitable for and accessible alike to professionals and interested laypeople.
> —Stephen Jay Gould, *Wonderful Life: The Burgess Shale and the Nature of History*

But just because people work for an institution they don't have to write like one. Institutions can be warmed up. Administrators and executives can be turned into human beings. Information can be imparted clearly and without pompous verbosity.
> —William Zinsser, *On Writing Well*

Much of what biologists write is for nonscientists. For example, biologists often write articles for newspapers and popular magazines, as well as essays for their students, colleagues, and peers. The secret to writing effectively for a general audience is to demystify science. Do this by connecting science to the everyday world and by showing readers that science is a human activity: an enterprise managed and mismanaged by men and women. As such, it involves luck and hunches. It also changes people's lives.

Magazines such as *Natural History, Discover*, and *Smithsonian* regularly publish articles written for nonscientists. These articles resemble and are often excellent models for term papers and essays. In these articles, biologists define terms, describe simple concepts, include images that readers understand, and discuss examples outside of science to communicate with readers. The writing is simple and logical, thereby helping convince readers that the conclusions are inevitable.

Analogies and examples, especially those with human interest, help a general audience understand what you're saying. Whatever the topic, you need to explain the unknown in terms of the known. Einstein was a master of this: He made his work accessible to nonscientists by writing about passing ships, flying birds, and other everyday experiences. For example, consider his definition of relativity: "When a man sits with a pretty girl for an hour, it seems like a minute. But let him sit on a hot stove for a minute, and it's longer than an hour. That's relativity."

Nonscientists have no use for the dull, jargon-laced writing of most scientists: They'll quickly pitch your writing aside if it doesn't meet their needs. You'll best attract the eye of nonscientists if you explain science in everyday language. For example, saying that "this substance has the consistency of pancake syrup" communicates ideas that would require many paragraphs to describe. By making science relevant, you make it meaningful to a general audience.

There are many ways to write about biology for a general audience. Whatever your approach, make sure that your paper has an introduction, a body, and a conclusion (Figure 6–1).

Introduction Introduce your paper in the first paragraph or two. Open with general statements that catch the reader's attention. Then progress to a focusing statement of the purpose of your paper, much as photographers begin their videos with panoramic shots before zeroing in on close-ups.

Body Develop your arguments in the body of your paper. The number of paragraphs in the body of your paper depends on the number of points that you want to make.

Conclusion Use the conclusion of your paper to tie everything together. Begin the conclusion of your paper by stating the paper's thesis. Widen the thesis to make a broad, concluding statement.

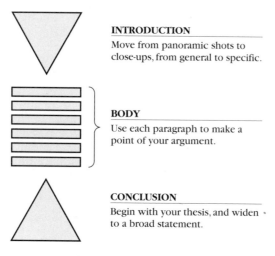

INTRODUCTION

Move from panoramic shots to close-ups, from general to specific.

BODY

Use each paragraph to make a point of your argument.

CONCLUSION

Begin with your thesis, and widen to a broad statement.

Figure 6–1

Model for the design of an essay.

Here are a few of the many ways to write for a general audience:

Analogies

Use analogies to link a scientific idea with a common experience. For example, analogies such as, "The eye is like a camera, with the retina being the film," communicate quickly and effectively with readers. Such analogies are easily developed into more detailed explanations:

> The human eye functions like a camera. The iris and the pupil of the eye function as the lens and shutter of the camera by controlling the depth of focus. The depth of focus depends on the size of the hole through which the photographer looks: The smaller the hole, the greater the depth of focus and the sharper the picture. And just like a camera's diaphragm, the pupil of the eye also shrinks or enlarges according to available light. When we squint, turn toward or away from bright light, or put on sunglasses, we mimic a photographer's adjustment of the camera's shutter-speed to let more or less light into the camera.

Consider how Richard Feynman uses numerical analogies to explain the significance of the magnitude of electrical forces:

> If you were standing at arm's length from someone and you had one percent more electrons than protons, the repelling force would be incredible. How great? Enough to lift the Empire State Building? No! To

lift Mount Everest? No! The repulsion would be enough to lift a weight equal to that of the entire earth.[1]

Here's how Paul and Anne Ehrlich use analogies (and sarcasm) to show the absurdity of the claim that "the world's food resources theoretically could feed 40 billion people":

> In one sense they were right. It's "theoretically possible" to feed 40 billion people—in the same sense that it's theoretically possible for your favorite major-league baseball team to win every single game for fifty straight seasons, or for you to play Russian roulette ten thousand times in a row with five out of six chambers loaded without blowing your brains out.[2]

Although such analogies communicate effectively when used properly, remember that they clarify, but do not prove, a point. If two things are alike in the two ways shown by the analogy, they're not necessarily alike in all ways.

Human Interest

This writer uses a "Look at me" opening to capture readers' attention long enough to lead them to read further:

> "My mother almost never drank," says Nancy Wexler, president of the Hereditary Diseases Foundation, "yet one day, as she was crossing the street, a policeman said, 'Aren't you ashamed to be drunk so early in the morning!'"
>
> Those words terrified her mother, for her father and her three older brothers had died of a frightening ailment called Huntington's disease, Wexler says, which made them lose their balance, twitch uncontrollably, and eventually lose their minds. Mrs. Wexler thought she had been spared: She was 53, past the age when symptoms of the disease usually begin. None of her family or friends had noticed the slight weave in her walk, for it had come about very gradually. But those startling and brutal words from the policeman sent her rushing to consult with her doctor.
>
> When the doctor confirmed that she was doomed to become totally incapacitated and die within 10 to 20 years, her husband, a Los Angeles psychoanalyst, called their two daughters home from graduate school. He had to explain not only that their mother had Huntington's but also that they, too, stood a 50–50 chance of getting the disease.[3]

Similarly, Robert Bazell opens his essay entitled "Biotech Bingo" in *The New Republic* with personal information that makes readers want to read on:

[1]Feynman, Richard P. 1964. *The Feynman Lectures on Physics.* Reading, MA: Addison-Wesley Publishing Co. II, 1.

[2]Ehrlich, Paul. R., and Anne H. Ehrlich. 1990. *The Population Explosion.* New York: Simon and Schuster, p. 19.

[3]Pines, Maya. 1984. "In the Shadow of Huntington's." *Science 84* 5: 32.

Last year Bill Smith's lymphoma, a cancer of the white blood cells, threatened to end his life within months. Horrendous open sores covered his skin. For years the disease had resisted several courses of standard chemotherapy and radiation treatments. So Smith volunteered for treatment at Boston University Hospital with an experimental drug made by Seragen, a small biotechnology firm in Hopkinton, Massachusetts. Doctors now find no trace of the cancer in Smith's body. His skin has cleared, and he says he feels great.[4]

This paragraph is also an excellent example of the importance of varied sentence-length and forceful concluding sentences.

News

In *National Geographic*, Esmond Martin starts his article with an interesting piece of news:

Sometime in the late 1970s half of the white rhinoceros population of Uganda suddenly disappeared—a single rhino, probably shot by a gang of poachers. At the time, nobody realized that it was one of only two left in the country.[5]

In this article from *Science News*, this writer saves the news for the last sentence of the paragraph:

In a mid-19th-century monastery garden, Gregor Mendel's experiments with smooth and wrinkled peas revealed the rules by which parents pass on traits to their offspring. But a theory called genomic imprinting is putting a new wrinkle into Mendelian genetics: A gene's expression may depend on which parent contributed it.
In the latest finding, . . .

Background Information

Other authors begin paragraphs with background information that leads to a question:

The sight of a flock of birds migrating south in the fall or north in the spring hardly ever fails to evoke a sense of wonder. The flight may be the orderly aerobatics of a V of Canada geese or the ragtag progress of a group of starlings. Whatever its details, the overwhelming impression

---

[4]Bazell, Robert. 1991. "Biotech Bingo." *The New Republic* 204: 10–12.
[5]Martin, Esmond B. 1984. "They're Killing Off the Rhino." *National Geographic* (March 1984): 404.

conveyed to the observer is that of a powerful inner impulse. The birds do not hesitate in their flight; they travel smoothly and unerringly toward a goal far out of the viewer's sight. Where does the impulse come from that guides the birds toward warmer climates in winter and brings them back to their northern breeding grounds in the spring?[6]

Every human cell contains oncogenes, genes that have the potential to cause cancer. These genes apparently carry out normal functions until a malignant change takes place. What is it that changes the oncogene from a normal part of the cell's genetic machinery into a source of cancerous, or neoplastic, transformation?[7]

Questions

Still others begin their papers with questions followed by answers. Here's how John Allen Paulos (author of *Innumeracy* and *Beyond Numeracy*) opened his "My Turn" column entitled "Orders of Magnitude" in *Newsweek*:

Quick: how fast does human hair grow in miles per hour? What is the volume of all the human blood in the world? If you don't know, it's no surprise; even math students sometimes don't, either. The answers are perhaps intriguing: hair grows a little faster than 10 to the minus-8th miles an hour (that is, a decimal point followed by 7 zeros and a 1); the totality of human blood will fill a cube 800 feet on a side, or cover New York's Central Park to a depth of about 20 feet.

Having gotten readers' attention with these questions and answers, Paulos then makes his point:

What does surprise me, however, is how often I find adults who have no ideas of easily imagined numbers: the population of the United States, say, or the approximate distance from the East Coast to the West. Many otherwise sophisticated people have no feel for magnitude, no grasp of large numbers like the federal deficit or small probabilities like the chances of ingesting cyanide-laced painkillers. This disability, which the computer scientist Douglas Hofstadter calls innumeracy, is so widespread that it can lead to bad public policies, poor personal decisions—even a susceptibility to pseudoscience.

---

[6]Gwinner, Eberhard. 1986. "Internal Rhythms in Bird Migration." *Scientific American* 254: 84.
[7]Croce, Carlo M., and George Klein. 1985. "Chromosome Translocations and Human Cancer." *Scientific American* 252: 54.

Contrast and Origin

William Brown combines contrast with origin to describe selection before 1900:

> For thousands of years Indians of the Western Hemisphere grew corn, varieties pollinated by the wind and bred largely by chance. Despite their lack of scientific insight, they transformed a wild grass from Mexico into one of the world's most productive plants. Sixteenth- and 17th-century farmers continued the practice of corn improvement. Distinct varieties were developed by selecting the best ears at the time of harvest and using seed from those ears to produce the next year's crop. This kind of selection continued until about 1900 and resulted in scores of high-yielding, randomly pollinated varieties. Then, in the course of just a few years, scientists applied genetics to corn breeding—and brought about a transformation of agriculture in this century.
>
> The development of hybrid corn resulted from the exploitation of a phenomenon known as heterosis or hybrid vigor. This increased yield, vigor, and rate of growth of plants comes from the mating of unrelated parents. Many early botanists and horticulturists, including Charles Darwin, had previously observed this phenomenon. But it was geneticist George Harrison Shull who developed the heterosis concept as it is applied today. He and E. M. East, a contemporary whose experiments at the Connecticut Agricultural Experiment Station in New Haven closely paralleled Shull's were the first to isolate pure strains of corn. These were then crossed to produce the reliable vigor of hybrid corn.[8]

Everyday Life

In this essay entitled "Following Aspirin's Trail," Steve Olson uses specific information about our medicine cabinets to grab our attention:

> Of all the drugs in a medicine cabinet, none is as familiar as aspirin. Most of us pop an aspirin as blithely as we take a shower or get a haircut. Americans swallow some 20 tons of the stuff everyday—enough to fill four good-sized dump trucks.
>
> But even though aspirin has been around since 1899, only in the last few years have scientists begun to uncover how it works. . . .

[8]Brown, William. 1984. "Twenty Discoveries That Changed Our Lives." *Science 84* 5: 65, 69, 73, 76, 79, 83, 99, 111, 115, 121, 127, 131–132, 141, 149, 153.

David Baltimore links our everyday lives to a less common bit of information—the discovery of the structure of DNA:

> Biology for all its mystery, is woven into the fabric of our daily lives. We cultivate plants and raise animals to eat; we burn wood to keep warm; we wear leather, cotton, or wool. Over the centuries, man has become increasingly adept at harnessing the biologic world, but before the 1950s our knowledge of how living things work was at best superficial. That has changed. One discovery stands out as the primary generator of our new understanding of biologic systems and our power to manipulate them: the 1953 elucidation of the structure of DNA by James D. Watson and Francis Crick.[9]

Several Points of View

In "A Charming Resistance to Parasites," Marlene Zuk states several points of view in the opening paragraphs:

> By 1889, British naturalist Alfred Russell Wallace had concurred with many of Darwin's ideas about evolution. He balked, however, at the notion that males of many animal species have evolved bright colors and complicated mating displays simply as a result of female's preference for them, and wrote:
>
> > . . . it may also be admitted . . . that the female is pleased or excited by the display. But it by no means follows that slight differences in the shape, pattern or colours of the ornamental plume are what lead a female to give the preference to one male over another. . . . A young man, when courting, brushes or curls his hair, and has his moustache, beard of whiskers in perfect order, and no doubt his sweetheart admires them; but this does not prove that she marries him on account of these ornaments, still less that hair, beard, whiskers and moustache were developed by the continued preferences of the female sex.
>
> Darwin disagreed and maintained that "the power to charm the female has sometimes been more important than the power to conquer other males in battle." Most evolutionary biologists would now agree with Darwin, but they still disagree over exactly what females find charming and how their tastes could cause the development of such bizarre

---

[9]Brown, William. 1984. "Twenty Discoveries That Changed Our Lives." *Science 84* 5: 65, 69, 73, 76, 79, 83, 99, 111, 115, 121, 127, 131–132, 141, 149, 153.

ornaments as a peacock's tail or the elongated wattles of a turkey. Recently, even the possible role of tiny blood parasites has been brought into the controversy.[10]

Misconceptions

Yves Dunant and Maurice Israel fight a misconception as they begin their paper in *Scientific American*:

> For many years there was a wide consensus among neuroscientists that acetylcholine is released from small, spherical organelles called synaptic vesicles, which are found inside the nerve terminal. It was thought that when the nerve terminal is stimulated, the vesicles fuse with the terminal membrane and release their contents into the space between the neuron and the tissue with which it communicates.
>
> Our recent investigations contradict this simple picture. They suggest that although the vesicles do indeed store acetylcholine and play a role in its regulation within the cell, the acetylcholine released by the nerve terminal does not originate in the vesicles. Instead the released acetylcholine is derived directly from the cytoplasm, which makes up the ground material inside the neuron. The releasing mechanism appears to be operated by a compound, most likely a protein, that is embedded in the membrane of the nerve cell. The protein may act as a valve, enabling acetylcholine to pass through membrane.[11]

This article entitled "Butterfly Fallacy" (from the April 11, 1991 issue of *USA Today*) presents the scientists' data to overturn a popular misconception about butterflies:

> For years, biology students have been taught that the delectable viceroy butterfly resembles its foul-tasting cousin, the monarch, to fool predators. Now David Ritland and Lincoln Brower of the University of Florida say the viceroy tastes lousy, too. Their study in the British journal *Nature* details how viceroys, monarchs and other butterflies known to be a good meal were stripped of their telltale wings and offered to 16 wild blackbirds. Only 41% of the viceroys were eaten vs. 98% of the tasty species. "It kind of shows how some of the obvious things we've assumed have never been tested," says Jim Miller, curator of moths and butterflies at the American Museum of Natural History in New York.

---

[10]Zuk, Marlene. 1984. "A Charming Resistance to Parasites." *Natural History* 93: 28.
[11]Dunant, Yves, and Maurice Israel. 1985. "The Release of Acetylcholine." *Scientific American* 252: 58.

Stating the Problem

These two authors concisely state the problem they studied:

> A seed falls, and finding the weather hospitable, it swells, splits its protective coat, and sends a shoot and root out into the air. Suddenly and without fanfare, the shoot curves upward and heads for the sky, while the root turns downward and angles its way into the soil.
>
> This uncanny ability to sense gravity and act accordingly, called gravitropism, has long fascinated botanists. But so far no one has been able to explain how plants get their directions straight.[12]

> The neuronal mechanisms of information storage remain one of the principal challenges in contemporary neurobiology. Over the past decade, however, the development of effective model systems has significantly advanced our understanding of the cellular basis of nonassociative learned behaviors. This progress has resulted largely from the exploitation of "simple" invertebrate models, but few effective systems are available for cellular analysis of associative learning or of learning in invertebrates.[13]

Descriptions

Biologists often describe their observations. The following excerpt from a National Cancer Institute publication uses a simple, straightforward style to describe the pancreas:

> The pancreas is a thin, lumpy gland about six inches long that lies behind the stomach. Its broad right end, called the head, fills the loop formed by the duodenum (the first part of the small intestine). The midsection of the pancreas is called the body, and the left end is called the tail.
>
> The pancreas produces two kinds of essential substances. Into the bloodstream it releases insulin, which regulates the amount of sugar in the blood. Into the duodenum it releases pancreatic juice containing enzymes that aid in the digestion of food.
>
> As pancreatic juice is formed, it flows through small ducts (tubes) into the main pancreatic duct that runs the full length of the gland. At the head of the pancreas, the main duct joins with the common bile duct, and together they pass through the wall of the duodenum. The common

---

[12]Fellman, Bruce. 1984. "How Do Plants Know Which End Is Up?" *Science 84* 5: 26.

[13]Gold, M.R., and G.H. Cohen. 1981. "Modifications of the Discharge of the Vagal Cardiac Neurons During Learned Heart Rate Change." *Science* 214: 345.

bile duct carries bile (a yellowish fluid that aids in the digestion of fat) from the liver and gallbladder to the duodenum.[14]

Interesting Phenomenon

Similarly, this author uses a simple writing style to describe an interesting phenomenon — how a snake smells with its tongue:

> For years it was thought that the snake's forked tongue was used to feel along in the dark, like the whiskers of a cat. Another popular belief was that it was poisonous. The forked tongue, which the snake can flick in and out without opening its mouth, through a gap in its lips, is actually an organ of smell.
>
> The tongue flicks out and waves around to attract a few odor molecules floating in the air. The two forks of the tongue are inserted briefly into two cavities in the top of the snake's mouth called Jacobson's organ. Sensors inside this organ immediately analyze the molecules and send the appropriate message to the snake's brain. The tongue can detect smells that signal danger, approaching prey, or the presence of a mate long before the snake can see or hear the source of the odors.[15]

Definitions

Biology and other sciences include many unusual terms, and writers must define these terms if they are to use them to communicate effectively with their readers. General audiences do not need a complete technical definition of a term to understand its meaning. Therefore, an adequate definition of penicillin for a general audience might be "a drug used to cure bacterial infections." The audience would be confused by this technically more accurate information presented in *Van Nostrand's Scientific Encyclopedia:*

> Penicillin is an antibacterial substance produced by microorganisms of the Penicillium chrysogenum group, principally penicillium notatum NRRL832 for deep or submerged fermentation and NRRL 1249, B21 for surface culture. Penicillin is antibacterial toward a large number of gram positive and some gram negative bacteria and is used in the treatment of a variety of infections.

---

[14]*What You Need to Know About Cancer of the Pancreas.* 1979. Washington, D.C.: National Cancer Institute.
[15]Thomas, Warren D., and Daniel Kaufman. 1990. *Dolphin Conferences, Elephant Midwives, and Other Astonishing Facts About Animals.* Los Angeles: Jeremy P. Tarcher, Inc.

Readability Formulas

Several people have tried to devise ways to measure the readability of writing. The most popular method is the "fog index" devised by Robert Gunning in 1944. The fog index of a paragraph or essay is the grade-level of education needed by a reader to understand the article. For example, an essay having a fog index of 12 could be read by a twelfth grader but not by a ninth grader. Similarly, a paragraph having a fog index of 16 could be understood by a college senior but probably not by a high school senior.

The fog index of a paragraph or essay is based on two important traits of writing: average sentence length and the percentage of difficult words. These features of writing must be matched to your audience's abilities if you expect your audience to read and understand what you've written. For example, most magazines have a consistent writing style:

Magazine	Average Sentence Length (words)	% Polysyllabic Words
True Confessions	12–15	3–4
Reader's Digest	14–17	8–9
Time	16–19	9–10

Other popular magazines such as *Newsweek, Harpers*, and *Ladies' Home Journal* all have average sentence lengths of under 22 words, and fewer than 12% polysyllabic words. This does not mean that all of the articles in these magazines sound alike; it merely means that the editor of each magazine matches the reading level of each article with the magazine's audience. Most scientific journals have a fog index that varies between 18 and 25, thus explaining why they are often so hard to read.

Here's how to determine the fog index of what you write:

- Choose a paragraph that you want to evaluate. Count the number of words and sentences in the paragraph.

- Divide the number of words by the number of sentences. This is the average sentence-length of the paragraph.

- Count the number of words having three or more syllables. Do not include proper names, combined words (for example, bookkeeper), or verbs created by adding endings such as *-ed* (for example, *created*).

- Divide this number of words by the total number of words in the paragraph. This is the percentage of polysyllabic words.

- Add the average sentence-length plus the percent polysyllabic words and multiply by 0.4. This is the fog-index of the writing.

As an example, consider this tongue-in-cheek description of some scientists' insistence on poor writing:

> It is the opinion of the writer that it is the appropriate moment to re-examine the style of writing which might most effectively be used by members of the biology profession. It is also the writer's belief that the long-lasting tradition about the inappropriateness of the active voice and the personal pronoun for technical writing has made for a great deal of inefficiency. This kind of writing has been exemplified in the past by numerous national publications. It would appear that an application of the principles of biology to the problem would be beneficial and it would seem the result might be that such a style would be eliminated.*

The numbers of words in each sentence of this paragraph are 31, 33, 14, 17, and 14. (Note that the last sentence has two independent clauses and is therefore counted as two sentences.) The total number of words in the paragraph (109) divided by the number of sentences (5) gives an average sentence length of 22 words. Furthermore, there are 20 words having three or more syllables. This means that 18% (20/109) of the words in the paragraph are polysyllablic.

The average sentence length plus the percentage of polysyllabic words is $22 + 18 = 40$. This number multiplied by 0.4 produces a fog index of 16, which corresponds to the reading level of a college senior. It is absurd to present the paragraph's simple idea at such a high level of reading difficulty. However, that effect was intentional. A translation of the paragraph reads like this:

> I think it's about time we stop insisting on impersonal style for biologists. I think that our national publications could set a good example in breaking this strait jacket. After all, the biologist wants efficiency and a "common sense" approach to his professional work. Why not encourage him to apply this practical method to his writing? If he does, he'll save time, money, material, and his reader's temper.

The fog index of this paragraph is 12.5. Not surprisingly, it's also easier to read and understand.

Readability formulas such as the one for determining the fog index can help you match your writing to your audience's ability to understand it. However, blindly following any formula or rule will not make you an effective writer. Moreover, formulas such as that for calculating a fog index cannot evaluate the precision or style of your writing. Use such formulas to check your work *after* you've written something, not to pattern or control your ideas before you write. Effective writing is lively: Don't kill it with this or any other "system."

*Edited from Shurter, Robert L. 1952. "Let's Take the Strait Jacket off Technical Style." *Mechanical Engineering*: 74: 664.

Notice how Acker and Hartsel, in their article entitled "Fleming's Lysozyme," effectively define difficult terms by first offering a common meaning:

> Bacterial anatomists are indebted to the late Sir Alexander Fleming for a sensitive chemical tool with which they have been studying bacteria, dissolving away the cell wall, and exposing the cell body or cytoplasm within. In 1922 at St. Mary's Hospital in London, six years before his epochal discovery of penicillin, Fleming found "a substance present in the tissues and secretions of the body, which is capable of rapidly dissolving certain bacteria." Because of its resemblance to enzymes and its capacity to dissolve, or lyse, the cells, he called it "lysozyme."[16]

You may want to define a term by discussing its origin. Here is an example:

> The word *biology* is derived from two Greek words, *bios*, "life" and *logos*, "word" or "discourse," and so "science." In its narrower sense, biology may be defined as the science of life, that is, the science which treats of the theories concerning the nature and origin of life; in a broader sense, biology is the sum of zoology (Greek *zoon*, "animal," +*logos*, "science"), or the science of animals, and botany (Greek *botone*, "plant"), or the science of plants. It is in this broad sense that the word is used in this volume.[17]

The first sentence of this paragraph gives the derivation of the word, and the second sentence provides details of the wider definition and its Greek origins. The final sentence is a transition between that definition and what will be discussed in the rest of the article.

There are many ways to write effectively for nonscientists. Choose the method best suited for your audience, purpose, and subject.

[16]Acker, Robert, and S.E. Hartsel. 1960. "Fleming's Lysozyme." *Scientific American*, June.
[17]Rice, Edward L. 1935. *An Introduction to Biology*. Boston: Ginn and Co.

James Watson

The Double Helix
"Finding the Secret
of Life"

James Watson (b. 1928) was a boy-wonder who
entered college when he was 15 and received a
Ph.D. in genetics when he was 20. Soon thereafter,
he teamed with Francis Crick to discover the
structure of DNA—work for which in 1962 they
received the Nobel Prize in Physiology and
Medicine. Watson described his view of that
collaboration and research in *The Double Helix*,
published in 1968. Throughout the book, Watson
mirrors himself as an arrogant, western cowboy.

Bragg was in Max's office when I rushed in the next day to blurt out
what I had learned. Francis was not yet in, for it was a Saturday
morning and he was still home in bed glancing at the *Nature* that had
come in the morning mail. Quickly I started to run through the details of
the B form, making a rough sketch to show the evidence that DNA was a
helix which repeated its pattern every 34 Å along the helical axis. Bragg
soon interrupted me with a question, and I knew my argument had got
across. I thus wasted no time in bringing up the problem of Linus, giving
the opinion that he was far too dangerous to be allowed a second crack
at DNA while the people on this side of the Atlantic sat on their hands.
After saying that I was going to ask a Cavendish machinist to make
models of the purines and pyrimidines, I remained silent, waiting for
Bragg's thoughts to congeal.

To my relief, Sir Lawrence not only made no objection but encouraged me to get on with the job of building models. He clearly was not in
sympathy with the internal squabbling at King's — especially when it
might allow Linus, of all people, to get the thrill of discovering the
structure of still another important molecule. Also aiding our cause was
my work on tobacco mosaic virus. It had given Bragg the impression that
I was on my own. Thus he could fall asleep that night untroubled by the
nightmare that he had given Crick carte blanche for another foray into

frenzied inconsiderateness. I then dashed down the stairs to the machine shop to warn them that I was about to draw up plans for models wanted within a week.

Shortly after I was back in our office, Francis strolled in to report that their last night's dinner party was a smashing success. Odile was positively enchanted with the French boy that my sister had brought along. A month previously Elizabeth had arrived for an indefinite stay on her way back to the States. Luckily I could both install her in Camille Prior's boarding house and arrange to take my evening meals there with Pop and her foreign girls. Thus in one blow Elizabeth had been saved from typical English digs, while I looked forward to a lessening of my stomach pains.

Also living at Pop's was Bertrand Fourcade, the most beautiful male, if not person, in Cambridge. Bertrand, then visiting for a few months to perfect his English, was not unconscious of his unusual beauty and so welcomed the companionship of a girl whose dress was not in shocking contrast with his well-cut clothes. As soon as I had mentioned that we knew the handsome foreigner, Odile expressed delight. She, like many Cambridge women, could not take her eyes off Bertrand whenever she spotted him walking down King's Parade or standing about looking very well-favored during the intermissions of plays at the amateur dramatic club. Elizabeth was thus given the task of seeing whether Bertrand would be free to join us for a meal with the Cricks at Portugal Place. The time finally arranged, however, had overlapped my visit to London. When I was watching Maurice meticulously finish all the food on his plate, Odile was admiring Bertrand's perfectly proportioned face as he spoke of his problems choosing among potential social engagements during his forthcoming summer on the Riviera.

This morning Francis saw that I did not have my usual interest in the French moneyed gentry. Instead, for a moment he feared that I was going to be unusually tiresome. Reporting that even a former birdwatcher could now solve DNA was not the way to greet a friend bearing a slight hangover. However, as soon as I revealed the B-pattern details, he knew I was not pulling his leg. Especially important was my insistence that the meridional reflection at 3.4 Å was much stronger than any other reflection. This could only mean that the 3.4 Å-thick purine and pyrimidine bases were stacked on top of each other in a direction perpendicular to the helical axis. In addition we could feel sure from both electron-microscope and X-ray evidence that the helix diameter was about 20 Å.

Francis, however, drew the line against accepting my assertion that the repeated finding of twoness in biological systems told us to build two-chain models. The way to get on, in his opinion, was to reject any argument which did not arise from the chemistry of nucleic-acid chains. Since the experimental evidence known to us could not yet distinguish between two- and three-chain models, he wanted to pay equal attention to both alternatives. Though I remained totally skeptical, I saw no reason to contest his words. I would of course start playing with two-chain models.

No serious models were built, however, for several days. Not only did we lack the purine and pyrimidine components, but we had never had the shop put together any phosphorus atoms. Since our machinist needed at least three days merely to turn out the more simple phosphorus atoms, I went back to Clare after lunch to hammer out the final draft of my genetics manuscript. Later, when I cycled over to Pop's for dinner, I found Bertrand and my sister talking to Peter Pauling, who the week before had charmed Pop into giving him dining rights. In contrast to Peter, who was complaining that the Perutzes had no right to keep Nina home on a Saturday night, Bertrand and Elizabeth looked pleased with themselves. They had just returned from motoring in a friend's Rolls to a celebrated country house near Bedford. Their host, an antiquarian architect, had never truckled under to modern civilization and kept his house free of gas and electricity. In all ways possible he maintained the life of an eighteenth-century squire, even to providing special walking sticks for his guests as they accompanied him around his grounds.

Dinner was hardly over before Bertrand whisked Elizabeth on to another party, leaving Peter and me at a loss for something to do. After first deciding to work on his hi-fi set, Peter came along with me to a film. This kept us in check until, as midnight approached, Peter held forth on how Lord Rothschild was avoiding his responsibility as a father by not inviting him to dinner with his daughter Sarah. I could not disagree, for if Peter moved into the fashionable world I might have a chance to escape acquiring a faculty-type wife.

Three days later the phosphorus atoms were ready, and I quickly strung together several short sections of the sugar-phosphate backbone. Then for a day and a half I tried to find a suitable two-chain model with the backbone in the center. All the possible models compatible with the B-form X-ray data, however, looked stereochemically even more unsatisfactory than our three-chained models of fifteen months before. So, seeing Francis absorbed by his thesis, I took off the afternoon to play tennis with Bertrand. After tea I returned to point out that it was lucky I found tennis more pleasing than model building. Francis, totally indifferent to the perfect spring day, immediately put down his pencil to point out that not only was DNA very important, but he could assure me that someday I would discover the unsatisfactory nature of outdoor games.

During dinner at Portugal Place I was back in a mood to worry about what was wrong. Though I kept insisting that we should keep the backbone in the center, I knew none of my reasons held water. Finally over coffee I admitted that my reluctance to place the bases inside partially arose from the suspicion that it would be possible to build an almost infinite number of models of this type. Then we would have the impossible task of deciding whether one was right. But the real stumbling block was the bases. As long as they were outside, we did not have to consider them. If they were pushed inside, the frightful problem existed of how to pack together two or more chains with irregular sequences of bases. Here Francis had to admit that he saw not the slightest ray of light. So when I walked up out of their basement dining room into the street, I left Francis with the impression that he would have to provide at least a semiplausible argument before I would seriously play about with base-centered models.

The next morning, however, as I took apart a particularly repulsive backbone-centered molecule, I decided that no harm could come from spending a few days building backbone-out models. This meant temporarily ignoring the bases, but in any case this had to happen since now another week was required before the shop could hand over the flat tin plates cut in the shapes of purines and pyrimidines.

There was no difficulty in twisting an externally situated backbone into a shape compatible with the X-ray evidence. In fact, both Francis and I had the impression that the most satisfactory angle of rotation between two adjacent bases was between 30 and 40 degrees. In contrast, an angle either twice as large or twice as small looked incompatible with the relevant bond angles. So if the backbone was on the outside, the crystallographic repeat of 34 Å had to represent the distance along the helical axis required for a complete rotation. At this stage Francis' interest began to perk up, and at increasing frequencies he would look up from his calculations to glance at the model. Nonetheless, neither of us had any hesitation in breaking off work for the weekend. There was a party at Trinity on Saturday night, and on Sunday Maurice was coming up to the Cricks' for a social visit arranged weeks before the arrival of the Pauling manuscript.

Maurice, however, was not allowed to forget DNA. Almost as soon as he arrived from the station, Francis started to probe him for fuller details of the B pattern. But by the end of lunch Francis knew no more than I had picked up the week before. Even the presence of Peter, saying he felt sure his father would soon spring into action, failed to ruffle Maurice's plans. Again he emphasized that he wanted to put off more model building until Rosy was gone, six weeks from then. Francis seized the occasion to ask Maurice whether he would mind if we started to play about with DNA models. When Maurice's slow answer emerged as no, he wouldn't mind, my pulse rate returned to normal. For even if the answer had been yes, our model building would have gone ahead.

The next few days saw Francis becoming increasingly agitated by my failure to stick close to the molecular models. It did not matter that before his tenish entrance I was usually in the lab. Almost every afternoon, knowing that I was on the tennis court, he would fretfully twist his head away from his work to see the polynucleotide backbone unattended. Moreover, after tea I would show up for only a few minutes of minor fiddling before dashing away to have sherry with the girls at Pop's. Francis' grumbles did not disturb me, however, because further refining of our latest backbone without a solution to the bases would not represent a real step forward.

I went ahead spending most evenings at the films, vaguely dreaming that any moment the answer would suddenly hit me. Occasionally my wild pursuit of the celluloid backfired, the worst occasion being an evening set aside for *Ecstasy*. Peter and I had both been too young to observe the original showings of Hedy Lamarr's romps in the nude, and so on the long-awaited night we collected Elizabeth and went up to the Rex. However, the only swimming scene left intact by the English censor was an inverted reflection from a pool of water. Before the film was half over we joined the violent booing of the disgusted undergraduates as the dubbed voices uttered words of uncontrolled passion.

Even during good films I found it almost impossible to forget the bases. The fact that we had at last produced a stereochemically reasonable configuration for the backbone was always in the back of my head. Moreover, there was no longer any fear that it would be incompatible with the experimental data. By then it had been checked out with Rosy's precise measurements. Rosy, of course, did not directly give us her data. For that matter, no one at King's realized they were in our hands. We came upon them because of Max's membership on a committee appointed by the Medical Research Council to look into the research activities of Randall's lab to coordinate Biophysics research within its laboratories. Since Randall wished to convince the outside committee that he had a productive research group, he had instructed his people to draw up a comprehensive summary of their accomplishments. In due time this was prepared in mimeograph form and sent routinely to all the committee members. The report was not confidential and so Max saw no reason not to give it to Francis and me. Quickly scanning its contents, Francis sensed with relief that following my return from King's I had correctly reported to him the essential features of the B pattern. Thus only minor modifications were necessary in our backbone configuration.

Generally, it was late in the evening after I got back to my rooms that I tried to puzzle out the mystery of the bases. Their formulas were written out in J. N. Davidson's little book *The Biochemistry of Nucleic Acids*, a copy of which I kept in Clare. So I could be sure that I had the correct structures when I drew tiny pictures of the bases on sheets of Cavendish notepaper. My aim was somehow to arrange the centrally located bases in such a way that the backbones on the outside were completely regular — that is, giving the sugar-phosphate groups of each nucleotide identical three-dimensional configurations. But each time I tried to come up with a solution I ran into the obstacle that the four bases each had a quite different shape. Moreover, there were many reasons to believe that the sequences of the bases of a given poly-nucleotide chain were very irregular. Thus, unless some very special trick existed, randomly twisting two polynucleotide chains around one another should result in a mess. In some places the bigger bases must touch each other, while in other regions, where the smaller bases would lie opposite each other, there must exist a gap or else their backbone regions must buckle in.

There was also the vexing problem of how the intertwined chains might be held together by hydrogen bonds between the bases. Though for over a year Francis and I had dismissed the possibility that bases formed regular hydrogen bonds, it was now obvious to me that we had done so incorrectly. The observation that one or more hydrogen atoms on each of the bases could move from one location to another (a tautomeric shift) had initially led us to conclude that all the possible tautomeric forms of a given base occurred in equal frequencies. But a recent rereading of J. M. Gulland's and D. O. Jordan's papers on the acid and base titrations of DNA made me finally appreciate the strength of their conclusion that a large fraction, if not all, of the bases formed hydrogen bonds to other bases. Even more important, these hydrogen bonds were present at very low DNA concentrations, strongly hinting that the bonds linked together bases in the same molecule. There was in addition the X-ray crystallographic result that each pure base so far examined formed as many irregular hydrogen bonds as stereochemically possible. Thus, conceivably the crux of the matter was a rule governing hydrogen bonding between bases.

My doodling of the bases on paper at first got nowhere, regardless of whether or not I had been to a film. Even the necessity to expunge *Ecstasy* from my mind did not lead to passable hydrogen bonds, and I fell asleep hoping that an undergraduate party the next afternoon at Downing would be full of pretty girls. But my expectations were dashed as soon as I arrived to spot a group of healthy hockey players and several pallid debutantes. Bertrand also instantly perceived he was out of place, and as we passed a polite interval before scooting out, I explained how I was racing Peter's father for the Nobel Prize.

Not until the middle of the next week, however, did a nontrivial idea emerge. It came while I was drawing the fused rings of adenine on paper. Suddenly I realized the potentially profound implications of a DNA structure in which the adenine residue formed hydrogen bonds similar to those found in crystals of pure adenine. If DNA was like this, each adenine residue would form two hydrogen bonds to an adenine residue related to it by a 180-degree rotation. Most important, two symmetrical hydrogen bonds could also hold together pairs of guanine, cytosine, or thymine. I thus started wondering whether each DNA molecule consisted of two chains with identical base sequences held together by hydrogen bonds between pairs of identical bases. There was the complication, however, that such a structure could not have a regular backbone, since the purines (adenine and guanine) and the pyrimidines (thymine and cytosine) have different shapes. The resulting backbone would have to show minor in-and-out buckles depending upon whether pairs of purines or pyrimidines were in the center.

Despite the messy backbone, my pulse began to race. If this was DNA, I should create a bombshell by announcing its discovery. The existence of two intertwined chains with identical base sequences could not be a chance matter. Instead it would strongly suggest that one chain in each molecule had at some earlier stage served as the template for the synthesis of the other chain. Under this scheme, gene replication starts with the separation of its two identical chains. Then two new daughter strands are made on the two parental templates, thereby forming two DNA molecules identical to the original molecule. Thus, the essential trick of gene replication could come from the requirement that each base in the newly synthesized chain always hydrogen-bonds to an identical base. That night, however, I could not see why the common tautomeric form of guanine would not hydrogen-bond to adenine. Likewise, several other pairing mistakes should also occur. But since there was no reason to rule out the participation of specific enzymes, I saw no need to be unduly disturbed. For example, there might exist an enzyme specific for adenine that caused adenine always to be inserted opposite an adenine residue on the template strands.

As the clock went past midnight I was becoming more and more pleased. There had been far too many days when Francis and I worried that the DNA structure might turn out to be superficially very dull, suggesting nothing about either its replication or its function in controlling cell biochemistry. But now, to my delight and amazement, the answer was turning out to be profoundly interesting. For over two hours I happily lay awake with pairs of adenine residues whirling in front of my closed eyes. Only for brief moments did the fear shoot through me that an idea this good could be wrong.

My scheme was torn to shreds by the following noon. Against me was the awkward chemical fact that I had chosen the wrong tautomeric forms of guanine and thymine. Before the disturbing truth came out, I had eaten a hurried breakfast at the Whim, then momentarily gone back to Clare to reply to a letter from Max Delbrück which reported that my manuscript on bacterial genetics looked unsound to the Cal Tech geneticists. Nevertheless, he would accede to my request that he send it to the *Proceedings of the National Academy*. In this way, I would still be young when I committed the folly of publishing a silly idea. Then I could sober up before my career was permanently fixed on a reckless course.

At first this message had its desired unsettling effect. But now, with my spirits soaring on the possibility that I had the self-duplicating structure, I reiterated my faith that I knew what happened when bacteria mated. Moreover, I could not refrain from adding a sentence saying that I had just devised a beautiful DNA structure which was completely different from Pauling's. For a few seconds I considered giving some details of what I was up to, but since I was in a rush I decided not to, quickly dropped the letter in the box, and dashed off to the lab.

The letter was not in the post for more than an hour before I knew that my claim was nonsense. I no sooner got to the office and began explaining my scheme than the American crystallographer Jerry Donohue protested that the idea would not work. The tautomeric forms I had copied out of Davidson's book were, in Jerry's opinion, incorrectly assigned. My immediate retort that several other texts also pictured guanine and thymine in the enol form cut no ice with Jerry. Happily he let out that for years organic chemists had been arbitrarily favoring particular tautomeric forms over their alternatives on only the flimsiest of grounds. In fact, organic-chemistry textbooks were littered with pictures of highly improbable tautomeric forms. The guanine picture I was thrusting toward his face was almost certainly bogus. All his chemical intuition told him that it would occur in the keto form. He was just as sure that thymine was also wrongly assigned an enol configuration. Again he strongly favored the keto alternative.

Jerry, however, did not give a foolproof reason for preferring the keto forms. He admitted that only one crystal structure bore on the problem. This was diketopiperazine, whose three-dimensional configuration had been carefully worked out in Pauling's lab several years before. Here there was no doubt that the keto form, not the enol, was present. Moreover, he felt sure that the quantum-mechanical arguments which showed why diketopiperazine has the keto form should also hold for guanine and thymine. I was thus firmly urged not to waste more time with my harebrained scheme.

Though my immediate reaction was to hope that Jerry was blowing hot air, I did not dismiss his criticism. Next to Linus himself, Jerry knew more about hydrogen bonds than anyone else in the world. Since for many years he had worked at Cal Tech on the crystal structures of small organic molecules, I couldn't kid myself that he did not grasp our problem. During the six months that he occupied a desk in our office, I had never heard him shooting off his mouth on subjects about which he knew nothing.

Thoroughly worried, I went back to my desk hoping that some gimmick might emerge to salvage the like-with-like idea. But it was obvious that the new assignments were its death blow. Shifting the hydrogen atoms to their keto locations made the size differences between the purines and pyrimidines even more important than would be the case if the enol forms existed. Only by the most special pleading could I imagine the polynucleotide backbone bending enough to accommodate irregular base sequences. Even this possibility vanished when Francis came in. He immediately realized that a like-with-like structure would give a 34 Å crystallographic repeat only if each chain had a complete rotation every 68 Å. But this would mean that the rotation angle between successive bases would be only 18 degrees, a value Francis believed was absolutely ruled out by his recent fiddling with the models. Also Francis did not like the fact that the structure gave no explanation for the Chargaff rules (adenine equals thymine, guanine equals cytosine). I, however, maintained my lukewarm response to Chargaff's data. So I welcomed the arrival of lunchtime, when Francis' cheerful prattle temporarily shifted my thoughts to why undergraduates could not satisfy *au pair* girls.

After lunch I was not anxious to return to work, for I was afraid that in trying to fit the keto forms into some new scheme I would run into a stone wall and have to face the fact that no regular hydrogen-bonding scheme was compatible with the X-ray evidence. As long as I remained outside gazing at the crocuses, hope could be maintained that some pretty base arrangement would fall out. Fortunately, when we walked upstairs, I found that I had an excuse to put off the crucial model-building step for at least several more hours. The metal purine and pyrimidine models, needed for systematically checking all the conceivable hydrogen-bonding possibilities, had not been finished on time. At least two more days were needed before they would be in our hands. This was much too long even for me to remain in limbo, so I spent the rest of the afternoon cutting accurate representations of the bases out of stiff cardboard. But by the time they were ready I realized that the answer must be put off till the next day. After dinner I was to join a group from Pop's at the theater.

When I got to our still empty office the following morning, I quickly cleared away the papers from my desk top so that I would have a large, flat surface on which to form pairs of bases held together by hydrogen bonds. Though I initially went back to my like-with-like prejudices, I saw all too well that they led nowhere. When Jerry came in I looked up, saw that it was not Francis, and began shifting the bases in and out of various other pairing possibilities. Suddenly I became aware that an adenine-thymine pair held together by two hydrogen bonds was identical in shape to a guanine-cytosine pair held together by at least two hydrogen bonds. All the hydrogen bonds seemed to form naturally; no fudging was required to make the two types of base pairs identical in shape. Quickly I called Jerry over to ask him whether this time he had any objection to my new base pairs.

When he said no, my morale skyrocketed, for I suspected that we now had the answer to the riddle of why the number of purine residues exactly equaled the number of pyrimidine residues. Two irregular sequences of bases could be regularly packed in the center of a helix if a purine always hydrogen-bonded to a pyrimidine. Furthermore, the hydrogen-bonding requirement meant that adenine would always pair with thymine, while guanine could pair only with cytosine. Chargaff's rules then suddenly stood out as a consequence of a double-helical structure for DNA. Even more exciting, this type of double helix suggested a replication scheme much more satisfactory than my briefly considered like-with-like pairing. Always pairing adenine with thymine and guanine with cytosine meant that the base sequences of the two intertwined chains were complementary to each other. Given the base sequence of one chain, that of its partner was automatically determined. Conceptually, it was thus very easy to visualize how a single chain could be the template for the synthesis of a chain with the complementary sequence.

Upon his arrival Francis did not get more than halfway through the door before I let loose that the answer to everything was in our hands. Though as a matter of principle he maintained skepticism for a few moments, the similarly shaped A-T and G-C pairs had their expected impact. His quickly pushing the bases together in a number of different ways did not reveal any other way to satisfy Chargaff's rules. A few minutes later he spotted the fact that the two glycosidic bonds (joining base and sugar) of each base pair were systematically related by a diad axis perpendicular to the helical axis. Thus, both pairs could be flipflopped over and still have their glycosidic bonds facing in the same direction. This had the important consequence that a given chain could contain both purines and pyrimidines. At the same time, it strongly suggested that the backbones of the two chains must run in opposite directions.

The question then became whether the A-T and G-C base pairs would easily fit the backbone configuration devised during the previous two weeks. At first glance this looked like a good bet, since I had left free in the center a large vacant area for the bases. However, we both knew that we would not be home until a complete model was built in which all the stereochemical contacts were satisfactory. There was also the obvious fact that the implications of its existence were far too important to risk crying wolf. Thus I felt slightly queasy when at lunch Francis winged into the Eagle to tell everyone within hearing distance that we had found the secret of life.

The Double Helix was written to be a best-seller, and it was. How would you describe Watson's writing style?

Exercises

1. In *The Double Helix*, James Watson writes about the relationship between meaning, style, and an author's personality:

    > By the time I was back in Copenhagen, the journal containing Linus' article had arrived from the States. I quickly read it and immediately reread it. Most of the language was above me, and so I could only get a general impression of his argument. I had no way of judging whether it made sense. The only thing I was sure of was that it was written with style. A few days later the next issue of the journal arrived, this time containing seven more Pauling articles. Again the language was dazzling and full of rhetorical tricks. One article started with the phrase, "Collagen is a very interesting protein." It inspired me to compose opening lines of the paper I would write about DNA, if I solved its structure. A sentence like "Genes are interesting to geneticists" would distinguish my way of thought from Pauling's.

    Write a one-sentence summary of this paragraph. What are the implications of Watson's message?

2. Watch an episode of a detective show such as *Columbo*. Write an essay describing how the detective used the scientific method to solve the case.

3. Write an essay summarizing a scientific or socio-scientific controversy such as the release of genetically altered organisms into the environment, societal control of science, or cloning as a means of human reproduction. Your audience for this essay is the layperson. Do not tangle yourself in the argument. Rather, discuss the issues and facts on which the controversy rests. Where facts are contradictory, say so, but do not try to resolve the argument.

4. In "Fibrinogen and Fibrin," Russell Doolittle uses these words to etch the concept of platelets into the reader's imagination:

    > Prick us and we bleed, but the bleeding stops; the blood clots. The sticky cell fragments called platelets clump at the site of the puncture, partially sealing the leak.[18]

    If Doolittle had written "Platelets help our blood to clot," he would have been accurate. Why, then, didn't he?

5. Select a biology-related article from a recent issue of *Scientific American* or *Natural History*. Assume that you are the journal's editor and must decide whether to publish the article. Write an essay summarizing your evaluation of the article and stating your recommendation regarding its publication.

6. Use an analogy to describe a molecule of ATP.

7. Describe the structure and function of a part of a cell to someone who knows no biology.

8. Find a paragraph in a textbook that you find hard to read, not because of substance, but because of expression. Why is the paragraph hard to understand? Rewrite the paragraph to improve its readability.

[18]Doolittle, Russell F. 1981. "Fibrinogen and Fibrin." *Scientific American*, December. p. 126.

9. List what you feel are the 20 most important ideas in science. Include a brief discussion of each idea and your ranking. Compare your ranking with that of Hazen and Trefil in *Science Matters: Achieving Scientific Literacy*:

> The universe is regular and predictable.
>
> One set of laws describes all motion.
>
> Energy is conserved.
>
> Energy always goes from more useful to less useful forms.
>
> Electricity and magnetism are two aspects of the same force.
>
> Everything is made of atoms.
>
> Everything—particles, energy, the rate of electron spin—comes in discrete units, and you can't measure anything without changing it.
>
> Atoms are bound together by electron "glue."
>
> The way a material behaves depends on how its atoms are arranged.
>
> Nuclear energy comes from the conversion of mass.
>
> Everything is really made of quarks and leptons.
>
> Stars live and die like everything else.
>
> The universe was born at a specific time in the past, and it has been expanding ever since.
>
> Every observer sees the same laws of nature.
>
> The surface of the earth is constantly changing, and no feature on the earth is permanent.
>
> Everything on the earth operates in cycles.
>
> All living things are made from cells, the chemical factories of life.
>
> All life is based on the same genetic code.
>
> All forms of life evolved by natural selection.
>
> All life is connected.

Do you agree with their ranking? Why or why not?

10. Calculate the fog indices of the Brown and Smith essays on page 4. What do these indices tell you? How do you explain these results relative to a survey indicating that scientists thought Smith was more dynamic and intelligent than Brown?

11. Discuss, support, or refute the ideas in these quotations:

> We, too, are silly enough to believe that all nature is intended for our benefit. —Bernard Le Bovier Sieur de Fontenelle
>
> Every cell from a cell. —Rudolph Virchow
>
> Every living thing from a living thing. —Louis Pasteur

Only time and money stand between us and knowing the composition of every gene in the human genome. — Francis Crick

Science when well digested is nothing but good sense and reason. — Stanislaw Leszczynski

Equipped with his five senses, man explores the universe around him and calls the adventure Science. — Edwin Powell Hubble

Science may be learned by rote, but Wisdom not. — Lawrence Sterne

CHAPTER SEVEN

Writing for a

Professional Audience

Only when he has published his ideas and findings
has the scientist made his scientific contribution,
and only when he has thus made it a part of the
public domain of science can he truly lay claim to it
as his. For his claim resides only in the recognition
accorded by peers in the social system of science
through reference to his work.
> —R.K. Merton

Science is organized knowledge.
> —Herbert Spencer

Doing an experiment is not more important than
writing.
> —E.G. Boring

Without publication, science is dead.
> —Gerald Piel

Work, finish, publish.
> —Michael Faraday

What Is a Research Paper?

The product of our work as scientists is not a table of data or a set of electron micrographs. Rather, it is an essay — a piece of expository prose called a research paper. A research paper is a written, published, and readily available report describing how the scientific method was used to study a problem. It is a repository of information — a permanent record describing repeatable experiments and original observations. It assesses observations, supports conclusions with data, evaluates conclusions, and describes the significance of the research. Research papers resemble a mystery with a solution documented by evidence. Moreover, they are the foundation of science. Indeed, 80% of all references in the scientific literature are to research papers in journals, and 85% of biomedical scientists rank research papers in journals as their most frequently used source of information for staying up-to-date.[1] To be published in a top-quality journal, a paper must be first judged worthy of publication by experts in the field. This so-called "peer review" improves the quality of published work and is a hallmark of first-rate journals.

The standard research paper was developed in the seventeenth century by Henry Oldenberg, secretary of the Royal Society of London.[2] Today, new journals appear every few months, and biologists' mailboxes regularly fill with announcements of new books and monographs. To stay up-to-date in their teaching and research, biologists must read a seemingly overwhelming number of books, papers, and journals.[3] This should help you better appreciate the proliferation of scientific journals:

In 1965 *Science Citation Index* surveyed 9,432 issues of 1,146 journals; in 1986 it surveyed 27,588 issues of 3,322 journals.

The journal *Biochemical and Biophysical Research Communications* originated in 1959. Volume 1 comprised 362 pages spread over 6 months; volume 106 comprised 1,489 pages spread over only 2 months.

Each year the *Journal of Immunology* publishes more than 5,000 pages.

[1]Levitan, Karen B. 1979. "Scientific Societies and Their Journals: Biomedical Scientists Assess the Relationship." *Social Studies of Science* 9: 393–400.

[2]*Philosophical Transactions*, the journal of the Royal Society of London, was founded by Oldenberg in 1665 and is the oldest scientific journal. Isaac Newton's discovery of the compound nature of sunlight, published in the February 19, 1672 issue of *Philosophical Transactions*, was the first major scientific discovery to be announced in a journal rather than in a book. Newton's paper was published 11 days after its receipt.

[3]A typical medical school library receives about 50 journals each day that require 61 cm (2 ft) of shelf space. Also, the number of scientific journals has doubled during the past decade. If this trend continues, in the year 2010 libraries will need 96,770 km (60,000 miles) of new shelf space each year for the new journals. Since 1960, the number of journals has increased by more than 2% per year, a bit slower than the growth in the number of scientists. This growth has occurred despite claims by scientists that few papers are innovative, that most papers are concerned only with details, and that the overall quality of research papers is poor. For more information about the proliferation of scientific journals, see Moran, Jeffrey B. 1989. "The Journal Glut: Scientific Publications Out of Control." *The Scientist* 10 July 1989, p. 11; Waksman, B.H. 1980. "Information Overload in Immunology: Possible Solutions to the Problem of Excessive Publication." *Journal of Immunology* 124: 1009–1015.

There are more than 30 monthly and bimonthly journals devoted exclusively or primarily to blood.

There are many kinds of journals, including those devoted to a specialty (*Journal of Lipid Research, Insect Biochemistry*), multiple disciplines (*Nature, Science*), multiple aspects of one science (*Biochemical Genetics*), one kind of organism (*Snakes, Birds*), one organ (*Heart, Brain*), one process (*Digestion*), and one disease (*Cancer*). Journals are published at different times: *Science* and *Nature* are published weekly; *Cell* biweekly; *Proceedings of the National Academy of Science* twice monthly; and *American Journal of Botany* monthly. Most biological papers in these journals are about six pages long and have about 15 references.[4]

Research papers are written by professional researchers and often include much technical language. Here are a few more examples of journals that publish biological research papers:

American Journal of Botany	*Biological Bulletin*
American Zoologist	*Developmental Biology*
Crop Science	*Ecology*
Ecological Monographs	*Heredity*
Evolution	*Journal of Bacteriology*
Immunology	*Journal of Cell Biology*
Journal of Biological Chemistry	*Journal of Morphology*
Journal of Experimental Biology	*Planta*
American Naturalist	*Annals of Botany*

If you want a career involving biological research, it's critical that you learn to write and publish research papers. Writing a paper for publication is largely an exercise in something that you studied in Chapter 2: organization.

How to Write a Research Paper

Science is a collabortive enterprise based on the open sharing of information that can be tested and built upon. The vehicle for sharing this information is the research paper published in a peer-reviewed journal such as *Ecology, Cell*, or *Plant Physiology*. Research is completed only when the results are published and made available to others.

---

[4]*Notes* and *Short Communications* are abridged forms of research papers.

Biologists can't read all the biological journals that are published. Even reading a few of these journals takes a tremendous amount of time and requires that scientists gather information efficiently.[5] To do this, scientists have developed a clear, standard format for reporting their discoveries. This prescribed format distinguishes scientific writing from literature and helps readers quickly understand what is being reported in the paper. Everything in a research paper must fit into one of the following categories:

Title

Authors and Affiliation

Abstract

Introduction

Materials and Methods

Results

Discussion

Literature Cited

Acknowledgments

These categories are not just names. Rather, they are guidelines that define the strategy and direction of your paper: They tell you what to put in and what to leave out of a paper. This structure has not resulted merely from chance or tradition. Indeed, it reflects the logic and elegance of the scientific method by sequentially progressing from the nature of the problem to the problem's solution. This uniform organization also helps scientists quickly and efficiently determine the relevance of papers they read to their work or interests. To publish a paper, and therefore to advance science, you must understand how scientific papers are organized.

Title

The title of a paper is similar to that of a newspaper article. It is a short label (usually less than 12 words) that helps readers decide if they will read your article. The title is your first chance to catch the reader's attention. Consequently, the title should indicate what the paper is about and thus ensure that your article is read by its intended audience. A title should be informative, specific, and concise and represent the content of the paper.

---

[5]Most scientists read only a few journals. For example, most medical researchers read about two weekly peer-reviewed journals, two monthly peer-reviewed journals, and two newsletters. Most biologists read about 10 journals per month. See Scheckler, W.E. 1982. "A Realistic Journal Reading Plan." *Journal of the American Medical Association* 248: 1987–1988; Berry, E.M. 1981. "The Evolution of Scientific and Medical Journals." *New England Journal of Medicine* 305: 400–402.

Thousands of biologists may read the title of your paper, either in the journal or in indexing publications such as *Current Contents*. Therefore, choose the title of your paper carefully. The title should identify precisely the main topic of your article. Trying to write an informative title is an effective test of whether your research has led to a definite conclusion. If you can't write a specific and informative title, you may not be sure of what you have to say. A vague title will cause your paper to miss its intended audience, thereby reducing the impact of your work.

Here's how to write an effective title for your paper:

Use the least number of words possible to adequately describe the content of your paper. Delete superfluous phrases such as, "Preliminary Studies of," "Aspects of," "Contributions to," and "Observations of" that tell the reader nothing. Similarly, avoid wordy and dull titles. "An Overview of the Structural and Conceptual Characteristics and Associated Psychometric Features of Future Planning Systems for Biological Research" strikes out on both counts. "Planning Biological Research" might get you to first base.

Make the title specific and informative. Avoid vague titles such as, "Enzymes in Plants." Such titles only raise more questions. What kinds of enzymes? What kinds of plants? What part of the plant? A more effective title would be, "Nitrate Reductase in Root Caps of Primary Roots of *Zea mays*."

Include taxonomic information where appropriate, and delete abbreviations, jargon, proprietary names, and chemical formulas.

Include the key words of the paper in the title.

If required to do so by the journal, provide a shortened title that will be used as a "running head" atop pages.

Not following these guidelines will lessen the impact of your paper. It may also produce a ridiculous title. For example, a paper published in *BioScience* titled "New Color Standard for Biologists" sounds like a taxonomic key for classifying purple, green, and orange biologists. Similarly, an article in the April 1991 issue of *BioScience* titled "Nematodes Win Prize at National Hardware Show" must have disappointed readers hoping that the prize would go to the raccoon's entry. My favorite titles from medical journals include "Stability of Prevalence" (What?), "Early Gastric Cancer in a United States Hospital" (Was the cancer in the lobby or in the kitchen?), and "Training Effect in Elderly Patients With Coronary Artery Disease on Beta Adrenergic Blocking Drugs" (Are trained drugs more effective than untrained drugs?). Finally, a paper published in *Clinical Research* was titled "Preliminary Canine and Clinical Evaluation of a New Antitumor Agent, Streptovitacin." When that pooch is through checking out streptovitacin, I hope it will drop by my lab to help me with my research.

Authors and Their Affiliations

The authors of a paper are the people who actively contributed to the design, execution, or analysis of the experiments and who take public intellectual responsibility for the paper. Authors are usually listed in the order of their importance to the work. The author listed first is the "senior author" and is the person who contributed most to the research. Similarly, the author listed last contributed least to the work.[6]

The "publish or perish" philosophy for promotion and tenure has increased the importance of authorship of scientific papers. The increased number of authors per paper reflects, in part, the increasing number of collaborations typical of modern science. However, authors are sometimes added to papers as a courtesy, thus explaining why the authorship of some papers reads like a laundry list or routing slip of a department or building. For example, a paper published in *Physical Review Letters* lists 27 authors yet is only 12 paragraphs long. Similarly, the first paper to describe a genetic linkage map had only one author,[7] while a more recent paper describing a linkage map of the human genome had 33 authors.[8] Sometimes, lab directors declare themselves an author despite not having done any of the research—a practice similar to the custom of "the medieval seigneur, whose prerogative it was to spend the marriage night with each new bride in his domain."[9]

Decide the authorship of a paper before you start writing the paper, even if the decision is tentative. Authors should know why and how the observations were made, how the conclusions were made from the observations, and how to defend the work against criticism. Furthermore, each author should have participated in the design, observation, and interpretation of the work. People who provide space, money, and routine technical assistance should be acknowledged for their contributions but usually should not be listed as authors of the paper. According to one set of guidelines for authorship of scientific papers, participation solely in the collection of data or other evidence does not justify authorship.[10] Most papers will have fewer than 4 authors, all of whom must take intellectual responsibility for the contents of the paper.

Following the list of authors are the authors' affiliations. If an author moves to another institution before the paper is published, indicate the new address with a footnote.

[6]Some journals have tried to avoid the ranking of authors by listing authors alphabetically. Interestingly, these journals received fewer papers from authors whose last names start with P–Z. See Garfield, E. 1977–1983. "More on the Ethics of Scientific Publication." In: *Essays of an Information Scientist*, Philadelphia: ISI Press. 5: 621–626.

[7]Sturtevant, A.H. 1913. "The Linear Arrangement of Six Sex-Linked Factors in *Drosophila*, as Shown by Their Mode of Association." *Journal of Experimental Zoology* 14: 43–59.

[8]Donis-Keller, H.P. Green, C. Helms, et al. 1987. "A Genetic Linkage Map of the Human Genome." *Cell* 51: 319–337.

[9]Chernin, E. 1981. "First Do No Harm." In: Warren, K.S., ed. *Coping with the Biomedical Literature.* New York: Praeger.

[10]Huth, E.J. 1986. "Guidelines on Authorship of Medical Papers." *Ann. Intern. Medicine* 104: 269–274.

Publish or Perish

Biologists at most major universities are under tremendous pressure to regularly publish the results of their research in peer-reviewed journals. The motive for this "publish or perish" pressure is that more papers presumably mean more prestige for a researcher's university and that prestige, administrators hope, will translate into more grant money for the university. This pressure to publish has many positive effects. For example, it encourages researchers to start projects that will generate publications that contribute to our understanding of life. However, the pressure to publish also has negative effects. At the very least, publish or perish encourages scientists to delay more important work to prepare for publication material that otherwise might not be submitted. At its worst, the pressure to publish promotes disreputable, unethical, and even fraudulent practices.

Many universities evaluate a scientist's productivity by counting the papers he or she has published. Not surprisingly, some scientists have used different strategies to increase their number of publications. Some have published the same paper more than once. This self-plagiarism indicates a lack of scientific objectivity, dilutes the literature, and pollutes the publication record of the scientist. A more common strategy has been to publish several small papers rather than one large paper. This "salami science" of slicing results into increasingly smaller pieces helps explain why almost half of the peer-reviewed research papers that appeared in the top 4,500 (of 74,000) scientific journals between 1981 and 1985 were never cited in the following five years.* Most of these uncited papers amount to little more than aggravating background noise in journals and are weak on objectivity, replicability, importance, competence, intelligibility, and efficiency.† These papers usually avoid important problems, do not challenge current beliefs, use complex methods, and are obtusely written to mask insignificance.

A final strategy to increase one's publication record has been to increase the number of authors on papers. When coauthors reciprocate the gratuity, they greatly increase their number of publications—all of the authors count the shared publications in their own "scores"—without increasing their workload. For example, the percentage of papers published with multiple authors in the *New England Journal of Medicine* has increased from 1.5% in 1886 (when the publication was called the *Boston Medical and Surgical Report*) to 96% in 1977. Some of this change is due to the shifting nature of scientific inquiry. However, much of the change is also due to the gratuitous inclusion of extra authors of publications.

Fortunately, many administrators and granting agencies have realized the problems associated with using number of publications to measure a scientist's worth. Agencies such as the National Science Foundation now require that scientists list no more than 3 publications per year (and 10 for a 5-year period) in their research proposals. This stimulates researchers to publish more full-length, thorough papers rather than numerous smaller reports.

*Institute of Scientific Information, as cited in Begley, Sharon. 1991. "Gridlock in the Labs." *Newsweek* 14 January 1991.
†Armstrong, J. Scott. 1982. "Research on Scientific Journals: Implications for Editors and Authors." *Journal of Forecasting* 1: 83–104.

Abstract

The abstract summarizes the major parts of the paper and is therefore a miniversion or skeleton of the paper. The abstract is important because it is a textual table of contents that maps your paper and helps readers assess its relevance to their interests. Abstracts are usually the first section of a paper read by reviewers (and readers of abstracting services such as *Biological Abstracts*).

Here are the features of an effective abstract:

It is one paragraph (usually shorter than 250 words) that summarizes the objective or scope of the paper, methods used in the work, results, conclusions, and significance of the research. It concisely states all of the paper's major results and conclusions. Notice how these authors stated their results in this abstract published in *Science*:

> Homing pigeons that had never seen the sun before noon could not use the sun compass in the morning; nevertheless they were homeward oriented.[11]

It is specific, concise, and must stand alone from the rest of the paper. Abstracts should include no abbreviations, acronyms, or citations of other papers.

The Abstract is not evaluative. Report only what's in your paper, and don't add your insights or thoughts.

Here's an example of a well-written abstract:

> Primary roots of *Zea mays* cv. Tx 5855 treated with fluridone respond strongly to gravity but have undetectable levels of abscisic acid. Primary roots of the carotenoid-deficient w-3, vp-5, and vp-7 mutants of *Z. mays* also respond strongly to gravity and have undetectable amounts of abscisic acid. Graviresponsive roots of untreated and wild-type seedlings contain 286 ± 31 and 317 ± 21 ng abscisic acid per gram fresh weight, respectively. We conclude that abscisic acid is not necessary for root gravitropism.

Although abstracts are relatively short, they are important. Well-written abstracts announce well-written papers, and poorly written abstracts usually forecast poorly written papers and poor science.

Introduction: What Was the Problem and Why Was the Work Done?

The introduction of a paper concisely states why you did the research and puts your work into the context of previous research.

[11]Wiltschko, R.D. Nohr, and W. Wiltschko. 1981. "Pigeons with a Deficient Sun Compass Use the Magnetic Compass." *Science* 214: 343.

Here's how to write an effective introduction:

Keep it short (two to five paragraphs) and simple. Do not write a comprehensive review of the literature: Include only enough information to help readers appreciate your work and to establish the background, context, and relevance of your research. Provide only enough background information to orient and help the reader determine how the research relates to what we do or don't know. Editors tire of reading comprehensive literature reviews and lengthy accounts of glorious accomplishments to which authors add their modest contribution in hopes of gaining glory by association. Long introductions waste time and space and are unnecessary, since review papers have eliminated the need for an extensive literature review in the Introduction of a paper. That doesn't mean that you can be ignorant of the history of the problem that you're

---

Verb Tense

You'll use present and past tense when you write a scientific paper. The work of others is part of an existing framework of knowledge; therefore, use present tense when you discuss the published work of others. Since new data are not yet considered part of that framework of knowledge, use past tense when reporting your own present findings. If you follow these guidelines, most of the Abstract, Materials and Methods, and Results sections will be written in past tense because they cite or describe your present work. For example:

Abstract
We studied the role of penicillin in preventing bacterial infections.

Materials and Methods
We fed our rats a diet containing 4% (w/v) oat bran.

Results
The experimental drug increased the size of nuclei in the cancerous cells.

Similarly, the Introduction and Discussion sections will be written mostly in present tense because they include frequent references to published research. For example:

Introduction
The availability of nitrogen strongly influences plant growth (Baldridge, 1992).

Discussion
Bass dominate the fish population of Lake Waco (Vodopich, 1992).

Here are the exceptions to these guidelines:

Use past tense when you refer directly to an author:
Baldridge (1992) studied leafhoppers at Marlin Prairie.

Use present tense when referring to a table:
Table 2 shows that . . .

studying. That knowledge will be revealed in your treatment of the data. There's no need to demonstrate that knowledge any other way.

Proceed from general to specific. Get quickly to the point to grab readers' attention. Do this with statements like, "This paper is the first to show that calcium affects osmoregulation in birds." Avoid statements such as "Alpha factor analysis has been shown to yield a lower bound estimate to the number of factors and allow psychometric interference to a universe of variables," which do little to introduce your paper. A likely response of readers to that sentence is, "Take me to your leader."

Don't tell readers what they already know and don't cite elementary textbooks. Rather, cite only significant papers that together show that a problem or question exists. Use these references to establish what is being challenged or developed and to show why you did the work.

Explain the rationale, objective, purpose, or hypothesis of the work. Tell readers why you did the work with statements such as "The objective of this research was to . . ." Define the problem, provide background information, and explain what your paper is about.

Don't force readers to read other papers to understand why you did the work. Use the introduction to set the tone of the paper. Do not use the introduction to try to convince readers of the importance of your work. Similarly, don't try to artificially inflate the importance of your work with mumbo-jumbo statements such as "The acquisitions of new observations makes it appropriate for us to re-evaluate our existing knowledge base and to make such modifications as are necessary to form a reasonably coherent theoretical whole." This pompous way of saying, "Knowledge begets knowledge" is silly and insulting to scientists having better things to do than read such gibberish.

Do not confuse a chronological review of the literature with an introduction to your research. Chronological approaches are usually ineffective because most biologists who read your paper are interested in and already know something about the topic. Starting your introduction with a nebulous and extensive rehash of the history of your field usually bores readers and prepares them for nothing. Avoid starting the Introduction of your paper with a sentence like this:

> The last decade has seen a rapid development of new techniques for studying phenomena associated with membranes.

Here are two typical examples of Introductions taken from research papers. Their important parts are labeled to show their order. I've abbreviated examples and omitted details where they seemed unnecessary. The identification of the part is enclosed in brackets; it applies to all of the material between it and the preceding bracket.

Although there have been various reports of multiple sclerosis (MS) in children (Low and Carter 1956, . . .), the onset of this disease occurs most frequently during adulthood and its manifestation during childhood is rare. [literature] For this reason the occurrence of MS in children often is not even considered, which may mean that the correct diagnosis is not made for young patients with a relapsing neurological disease, although the clinical condition may be indicative. [problem]

A diagnosis of MS basically depends on clinical criteria, i.e., neurological signs suggesting multiple lesions of the nervous system and a relapsing course (Poser et al. 1983). Findings from laboratory investigations, such as delayed latencies of evoked responses and the presence of oligoclonal antibodies in the cerebrospinal fluid (CSF), may lend diagnostic support to clinically suspected MS. Moreover, the detection of cerebral plaques by means of computer tomography (CT) is of particular importance and is considered to be of pathognomonic value (Bye et al. 1985). [discussion of past methods] However, there is no specific test to confirm this diagnosis. [problem] Since the introduction of magnetic resonance imaging (MRI), it has been expected that this method would increase diagnostic certainty in cases of suspected MS. Previous reports on MRI findings for adults suffering from MS confirm this expectation (Gebarski et al. 1985). . . . [discussion and literature]

In the present paper we report and contrast MRI and CT findings for three children who were diagnosed as definite MS cases according to the criteria of Poser et al. (1983) and who were treated in our hospital between 1980 and 1985. [objective, methods]

Since 1975, ion transport across dog tracheal epithelium has been studied extensively by using tissue sheets mounted in Ussing chambers (20). [literature] In such a preparation, the basolateral membranes of the cells are separated from the bathing medium by a layer of connective tissue 500–1,000 μm thick. [past method] The size of this collagenous dead space makes it impractical to study the exchange of ions across the basolateral side of the tissue. [problem] To overcome this problem, we decided to develop an isolated cell preparation in which both apical and basolateral membranes would be in direct contact with the bathing medium. [objective, new method] Such a preparation should allow us to study how Na and Cl enter the cells from the serosal side of the tissue. [advantage of new method]

This paper describes a method for obtaining dispersed isolated cells, the viability of which is assessed by a number of methods. [subject of paper]

Clearly isolated cell preparations offer several advantages, and our preparation will, we hope, prove suitable for a variety of studies on the biochemistry of the tracheal mucosa. [importance]

Heed the advice of an editor of *Nature*, a prestigious scientific journal: "If more authors . . . would attempt to make the first paragraph into a crystal clear description of what the paper is about rather than what other people's papers have been about, *Nature* would be an easier journal to read."[12]

Materials and Methods: How Did I Study the Problem? What Did I Do?

The Materials and Methods section describes how, when, where, and what you did. The hallmark of the Materials and Methods section of a paper, and of the scientific method, is repeatability, which, in turn, is determined by how well you write the Materials and Methods section. If a competent biologist with a similar background can repeat your work and obtain similar results, the Materials and Methods section is well written. If the materials and methods are not repeatable, this represents poor science, and the paper should be rejected, regardless of its results or conclusions. Write the Materials and Methods section while you're doing the work, when your ideas are fresh.

Here's how to write an effective Materials and Methods section:

Materials

Describe the growth conditions, chemicals, lighting, temperature, diet, and apparatus used in the work. List sources of hard-to-find materials and avoid mentioning brand names unless they are critical for repeating the work. Provide diagrams or photographs of unusual set-ups or equipment. Be as specific as possible; for example, say "methanol" instead of "alcohol."

Include the genus and species (e.g., *Zea mays*, not corn), strain or cultivar, characteristics, age, and source of all organisms. In studies involving humans, indicate how the subjects were selected. If required by the journal, attach an "informed consent" statement to the manuscript (Appendix 7).

Italicize or underline the scientific name of the organism(s) that you studied. Capitalize the genus and write the species in lowercase letters.

human   *Homo sapiens* or Homo sapiens

corn   *Zea mays* or Zea mays

Capitalize larger divisions such as phyla, classes, orders, and families.

Chordata   Primates

Capitalize proper names that are part of a medical term.

Hodgkin's disease

---

[12]Editorial. 1975. "It's Your Journal." *Nature* 252: 337.

Write the names of elements in lowercase letters. Capitalize the first letter of their symbols.

sodium   Na

iron   Fe

Describe field sites with photos or maps.

Methods

List methods chronologically with subheadings such as "Microscopy," "Sampling Techniques," and "Statistical Analyses." The Materials and Methods section is usually the first section of the paper to have subheadings.

Describe all controls and variables that you tested. Describe features such as temperature, pH, photoperiod, and incubation conditions. Do not include information such as the day the experiments were done or the type of microscope you used unless this information is critical to repeating the experiment. State the sample sizes and describe statistical treatments (see Chapter 8).

Reference previously published methods only if they appear in widely available journals. Provide all information about new equipment, new techniques, modified techniques, or techniques described in obscure journals. Avoid citing obscure journals such as *The Indonesian Journal of Zebra Nostril Hairs*. Chances are that most libraries won't have it.

Here's an example of a typical Materials and Methods section of a paper published in *Heredity*. This paragraph describes how the researchers digested DNA with restriction endonucleases:

DNA was digested with restriction endonucleases at 37°C in conditions recommended by the supplier (Bethesda Research Laboratories). Eight restriction endonucleases all specific for different hexanucleotide sequences were used. The enzymes were BamHI, EcoRI, HindIII, Hpa-I, Pst-I, Sal-I, Sba-I, Xho-I. Electrophoresis of single and double endonuclease digests was carried out in 1 per cent agarose gels in a BRL, H4 gel apparatus. The continuously circulated buffer was standard $1 \times$ TBE; Tris (89 mM)-boric acid (89 mM)-EDTA (2-5 mM) and low voltage (1–5 volts cm^{-1}) gradients were used (McDonnell, Simon and Studier, 1977) to separate the DNA fragments.

Write the Materials and Methods section as soon as you've established your procedures and have overcome any initial problems. Keep the reader in mind; for example, ask yourself if the reader is familiar with and can reproduce the technique. If so, use references to provide details of what you did and to avoid long recipes. Remember that the issue is repeatability and that evidence is invalid without repeatability.

Laboratory and Field Notebooks

When found, make a note of. —Charles Dickens

Laboratory and field notebooks are important tools of scientists. These notebooks are a cross between a journal and a class notebook and contain a complete record of your work, observations, and ideas. Biologists use them to (1) record data about laboratory and field observations; (2) record background information relevant to their experiments; (3) evaluate, plan, and describe experiments; and (4) speculate about results. These notebooks are one of the few places where you can openly question your data.

All entries in notebooks should be complete, permanent, efficient, and systematic recordings of observations and include the date, time, and conditions of the observations. Recording information and ideas as you go will reveal gaps in your knowledge and help you understand what you did. This, in turn, will help you better understand and prepare for future experiments. Notebooks also minimize your reliance on memory or random scraps of paper with cryptic notes. Document variations and details of your work, for this information could be important for later discoveries. Insert evidence such as photographs, sketches, and print-outs to help yourself formulate ideas.

Having a thorough, accurate, and well-organized notebook will greatly ease writing, because the notebook will contain ideas and information that will be included in the report.

Results: What Were the Findings?

The Results section of a paper reports new knowledge and is the heart of a scientific paper. The early parts of a paper describe how you got to the Results, and the later parts describe what the results mean. Thus, the entire paper stands on the Results section.

The Results section should summarize your findings and describe the evidence for your arguments. It is often the shortest section of a paper. Here's how to write an effective Results section:

Present relevant, representative data from your experiment, not all data from all experiments that you might have done. Summarize your data to show trends and patterns.

Make no comparisons with data in other publications, and do not discuss why your results agree or disagree with your predictions. Do that in the Discussion section. When making comparisons, avoid bloopers such as, "Cases also smoked significantly more cigarettes than controls . . ." My cases smoke only when the record room is on fire.

Present repetitive determinations in tables or graphs (see Chapter 8). Use graphs and tables to support generalizations about your data.

Avoid redundancy. Present data in only one way: Do not use text to describe data presented in a graph or table. Refer to all tables and figures in the text.

Guide the reader and point out trends. Avoid arrogant, opinionated statements such, "Our data clearly show that. . . ." Your readers, not you, will decide if your data show anything, much less if they show them clearly.

Make it simple and direct. For example, write that "penicillin inhibited growth of *E. coli* (Table 1)." Similarly, don't say that ". . . the patient experienced a rapidly fatal outcome" if all you mean is that the patient died.

Be sure that all methods used to obtain the results are described in the Materials and Methods section of the paper.

Support conclusions drawn from numerical data with brief statements of statistical criteria that you used.

Here's an example of a typical Results section published in the *Journal of Environmental Quality:*

> In the first experiment, the highest acidity (pH 2.6) significantly reduced the mass of hypocotyls, but there were no significant effects of acidity on shoots (Table 2). There also were no significant effects of anions on the mass of either shoots or hypocotyls, nor was there an interaction between acidity and anions. Both linear and quadratic terms in the dose-response functions for effects of acidity on hypocotyls were significant.
>
> In the second experiment, simulated rain at pH 3.0 and 3.4 reduced the mass of hypocotyls compared to pH 5.0, but there were no significant effects of acidity on shoots (Table 3). Anion composition of simulated rain did not significantly affect mass of shoots or hypocotyls nor was there a significant interaction between acidity and anions. The same results were found in harvests 1 and 2 although the effects of acidity on mass of shoots was close to being significant at $\alpha = 0.05$ in harvest 1.
>
> Generally, hypocotyls were more susceptible to effects of acidity than shoots (Tables 2 and 3) and effects tended to become less pronounced when plants were given a recovery period after treatment (harvest 2, Table 3). Daily exposure of simulated acidic rain during the period of rapid hypocotyl expansion appeared to have a slightly greater effect than three exposures per week beginning with the seedling stage (Table 2 vs. Table 3). Anions had no effect on dry mass, alone or in combination with acidity, either in the first or second experiment.

Discussion: Why Are the Findings Important?

The Discussion section tells what the results mean and why they're important. It interprets the results relative to the objectives stated in the Introduction and answers the questions, "So what?" and "What does it mean?" The Discussion section is where the true nature of the paper comes to light. It is usually the most difficult section of a paper to write.

Here's how to write an effective discussion section:

Tell the reader what the results mean. For example, do they support the hypothesis you tested? Why or why not? Don't leave your readers wishing that you would explain your explanation.

Write simply, clearly, and concisely. Faulty logic is much easier to cover up if you allow yourself to write poorly.

Clearly discuss generalizations, relationships, principles, and the significance of your results. Deliver on what you promised in the Introduction, and do not merely restate your results. Discuss and analyze your data, not the methods or statistics.

Summarize evidence for each of your conclusions. However, avoid obvious conclusions such as these from the medical literature:
". . . unproductive diagnostic measures are unnecessary." Agreed.
"the common practice of misdiagnosing deep vein thrombosis clinically should be abandoned." Agreed.
"Very obviously, mouse connective tissue is not necessarily human connective tissue." Very obviously.

Use your data and those of others to argue for the most plausible interpretation for your results. Propose a simple, testable hypothesis to explain your results. Remember to keep it simple; piling hypotheses atop each other is bad for a reader's digestion and an author's reputation. Also remember that it's not enough to collect or report data—machines can do that. Rather, you must explain the *significance* of your results.

Focus on important discoveries and their underlying causes. Do not present all conceivable explanations of your data. It's usually OK to speculate if the speculation is reasonable, based on data, and testable. Use data to support your arguments.

Don't overstate your data, and remember that you can't show the "whole truth;" leave that to the loudmouths who proclaim it every day. Research is often incomplete yet valuable because it points to new experiments or new ways of organizing information. Darwin recognized the incomplete nature of his work in the Introduction to *The Origin of Species:*

> No one ought to feel surprise at much remaining as yet unexplained in regard to the origin of species and varieties, if he make due allowance for our profound ignorance in regard to the mutual relations of the many beings which live around us. . . . Although

much remains obscure, and will long remain obscure . . . I am convinced that natural selection has been the most important, but not the exclusive, means of modification.

Darwin left many findings obscure or unexplained. Nevertheless, his ideas are among the most important ideas ever proposed. Darwin's work redirected the ideas of other scientists and led to much important work, particularly in areas in which his theory was weak. Do not discount the importance of interim solutions.

Move from specifics to generalizations, and compare your findings with those of others. Point out exceptions, lack of correlation, and unexpected results. Seek explanations, not refutations, and never state the opinion of others or the majority as fact.

Discuss the implications and importance of your work. Remember that data do not suggest, research doesn't indicate, and results do not show: These are all actions of scientists, not of data. Show your knowledge and take responsibility for your work by refusing to hedge, apologize or retreat behind a wall of excuses and "hedge words." Take a stand and finish the section positively and forcefully with statements such as, "I conclude that. . . ." Then stop.

Acknowledgments

In most of mankind gratitude is merely a secret hope for greater favours.—Duc de la Rochefoucauld

The Acknowledgments section of a paper is where authors acknowledge organizations, reviewers, and colleagues who helped with the paper or the research. The most common acknowledgments are for technical help, the use of equipment, and financial support. Avoid flowery phrases such as "magnanimous generosity" and "who contributed unselfishly." Instead, just say something like "I thank Randy Wayne for his help with the micromanipulators." Also avoid awkward acknowledgments such as this one: "The death of Dr. A.P. Meiklejohn, who helped in the preparation of the first edition, must have added greatly to the work of the authors and they are to be congratulated on their efforts."

Acknowledgments are a professional courtesy and do not imply an endorsement of the work. Many scientists expect colleagues to acknowledge their help, while others do not want to be acknowledged. Therefore, check with your colleagues before acknowledging them in your paper.

Literature Cited

. . . the function of a citation is not different from that of the paper itself: to supply the reader with information he doesn't already have.—E. Garfield

The man is most original who can adapt from the greatest number of sources.—Thomas Carlyle

The Literature Cited section of a paper provides complete bibliographic details of all work cited in the paper. As such, this section substantiates many claims made in the paper and is where you document evidence provided in other papers. Here's how to write an effective Literature Cited section:

Include all required details of all papers cited in the paper. Carefully check the accuracy of each citation. Interestingly, this section of a scientific paper is where most mistakes are made: Authors regularly include misspelled names, added or deleted names, incorrect dates, mangled titles (even of papers written by the author), and wrong journals.[13] The carelessness that produces these errors can damage your scientific credibility, for not checking the accuracy of supporting data is like making conclusions about a chemical reaction without checking the purity of the starting products. Examine all of the original papers yourself so that you won't perpetuate others' mistakes.

Cite only significant, published papers. Increasing the number of literature citations changes neither your data nor your conclusions; majority does not equal fact or truth. As William Roberts said, "Manuscripts containing innumerable references are more likely a sign of insecurity than a mark of scholarship."

Build the Literature Cited section as you write the paper. Cite the references in the format specified by the journal in its "Instructions to Authors." Although different journals require different formats for literature citations, three systems dominate the scientific literature: the Harvard System, the Alphabetical System, and the Citation Order System.

The **Harvard System** lists names of authors in the text: ". . . according to procedures described by Smith and Wesson (1990)." This system is easy to use when writing and allows readers to recognize authors of papers that you cite. *Heredity, American Journal of Botany, The Journal of Cell Biology, Evolution, Ecology,* and *Cell* are journals that use the Harvard System of citing literature. Here's an example of a citation:

Several biologists have suggested that dinosaurs were warm-blooded (Flintstone and Rubble, 1991).

The **Alphabetical System** refers to papers by number: ". . . according to procedures described by Smith and Wesson (1)." Cited papers are listed alphabetically in the Literature Cited section of the paper. *Plant*

[13]Poyer, R.K. 1979. "Inaccurate References in Significant Journals of Sciences." *Bulletin of the Medical Librarians Association* 67: 396–398; Key, J.D., and C.G. Roland. 1977. Reference Accuracy in Articles Accepted for Publication in the Archives of Physical Medicine and Rehabilitation." *Archives of Physical Medicine* 58: 136–137.

Copyright and Ownership

You may want to include quotations, illustrations, and photographs from other sources, including your own, in your papers. If your paper is used for educational purposes, involves no profits, and does not significantly impact the original paper, you can probably use the material without worrying about copyright problems. For example, including short quotations from other papers does not require that you obtain any permissions, because such use is considered "fair use" by the government. However, if you want to use large quotations, photographs, or other illustrations from other papers in your paper, you must know something about copyright.

Copyright is a legal protection against unauthorized use of your work. Publication or registration with the U.S. Copyright Office is not required for a work to be protected by copyright: the copyright belongs to a writer or artist as soon as he or she finishes writing a paper, book, or essay. This copyright provides exclusive rights to reproduce copies, derive works from the work, rent or sell the work, and transfer the copyright to someone else.

Most publishers require that you transfer the copyright of your work to them to protect their interests and to prevent unauthorized use of your paper. As dictated by the U.S. Copyright Act of 1976, this transfer must be in writing on a copyright transfer form. This transfer helps ensure that your paper is original, published only once, and that everyone, including yourself, must obtain written permission to use data or photographs included in your paper. If you want to republish anyone's work, seek the author's early. Request from the holder of the copyright exactly what's wanted; for example, send a photocopy of the table or figure that you want to republish. It is the job of the author, not the editor, to obtain permission to republish copyrighted information. Do this by sending a "permission request letter" (Appendix 4) to the copyright holder and author. The copyright holder will stipulate what to include in the credit line and whether a fee is required.

Transfer of copyright gives to someone else the right to use the article, not the article itself. Such a transfer is necessary to maintain the integrity of scientific publications because it allows publishers to protect your work from unauthorized use. It also means that you'll need written permission from the publisher to republish tables, figures, or substantial parts of the text of your paper. Publishers will honor all reasonable requests to publish your work in review articles, collections of reprints, or institutional collections of papers. When you republish this material, indicate that you've obtained written permission to reprint material with a caption such as, "Reprinted with permission from (source); copyright (year) by (owner of copyright)." There's no need for you to obtain permission to use material in the "public domain," which includes works whose copyright has expired and works prepared by officers or employees of the U.S. government as part of their official duties.

If you have questions about copyright, contact the U.S. Copyright Office, Library of Congress, Washington, DC 20559.

Physiology and *Journal of Cell Biology* are journals that use the alphabetical system of citing literature. Here's an example of a citation:

> Several biologists have suggested that dinosaurs were warm-blooded (1).

The **Citation Order System** also refers to papers by number. References are arranged according to the order in which they appear in the text. *Science, Journal of Sports Medicine and Physical Fitness,* and *Biopolymers* are examples of journals that use the citation order system of citing literature. Here's an example of a citation:

> Several biologists have suggested that dinosaurs were warm-blooded (8).

Unless required to do otherwise by the journal, cite the paper where the reference best applies in the sentence. For example, the sentence, "We used previously published methods to measure the protein content and respiratory rate (Doe, 1986; Smith, 1990)" does not specify what reference matches each technique. The sentence is improved by placing the references where they best apply: "We used previously published methods to measure the protein content (Doe, 1986) and respiratory rate (Smith, 1990) of the rats."

The "Instructions to Authors" will include directions for handling other problems associated with citing literature, such as citing papers having more than one author and citing papers published by the same authors in the same year.

Almost all journals use different formats to list cited papers. Here are a few examples:

Journal of Cell Biology
Moulder, J.W. 1985. Comparative biology of intracelluar parasitism. *Microbiol. Rev.* 49:298–337.

Evolution
Turelli, M. 1988. Phenotypic evolution, constant covariances, and the maintenance of additive variance. Evolution 42:1342–1347.

Genetics
Snadler, L., D.L. Lindsley, B. Nicoletti and G. Trippa, 1968. Mutants affecting meiosis in natural populations of *Drosophila melanogaster.* Genetics 60: 525–558.

Ecology
Kareiva, P.M. 1987. Habitat fragmentation and the stability of predator–prey interactions. Nature 326: 388–390.

Cell
Russell, J.H. (1983) Internal disintegration model of cytotoxic lymphocyte-induced target damage. Immunol. Rev. *72,* 92–118.

Journal of Protozoology
Smetacek, V. 1981. The annual cycle of protozooplankton in the Kiel Bight. *Mar. Biol.*, 63:1–11.

Journal of Insect Physiology
Hillman, W.S. (1973) Non-circadian photoperiodic timing in the aphid, *Megoura viciae. Nature 242,* 128–129.

Journal of Mammalogy
Van Horne, B. 1983. Density as a misleading indicator of habitat quality. The Journal of Wildlife Management, 47:893–901.

Animal Behaviour
Jackson, J.A. 1977. Red-cockaded woodpeckers and pine red heart disease. *Auk*, 94, 160–163.

Science
T.D. Gilmore, *Cell* 62, 841 (1990).

American Journal of Botany
Spurr, A.R. 1969. A low-viscosity epoxy resin embedding medium for electron microscopy. *Journal of Ultrastructure Research* 26: 31–43.

Many journals require authors to list the entire name of a journal, while others require only the abbreviation of the journal's title. If you use abbreviations, don't abbreviate one-word titles of journals (e.g., *Science*). The word "Journal" is abbreviated "J." and "-ology" words are cut off after the "l" (for example, *J. Bacteriol.* for *Journal of Bacteriology*).

Choosing a Journal

In almost every scientific subfield there is a hierarchy of journals that reflects the relative quality of published papers. Although it does not exist overtly, this hierarchy is known to all sophisticated scientists within the field.
   —National Academy of Sciences, *The Life Sciences*

Decide to what journal you'll submit your paper before you begin writing. Use these criteria to decide where to submit your paper:

Scope of the Journal   Each journal publishes a statement of its purpose. For example, here's what *Evolution* tells its authors:

Manuscripts submitted to *Evolution* should contain significant new results of empirical or theoretical investigations concerning facts, processes, mechanics, or concepts of evolutionary phenomena. Brief notes, or comments on previously published papers, should be submitted as "Notes and Comments" . . . Acceptance is based upon the significance of the article to the understanding of evolution; each paper must stand on its own merits and be a substantial contribution to the field.

Send your paper to a journal whose purpose matches the objectives of your paper.

Prestige of the Journal   A journal's prestige results largely from the quality of papers that it publishes. Consequently, your paper will have the greatest impact if it is published in a prestigious multidisciplinary journal such as *Science* rather than in either a low-quality journal or a journal not read by your colleagues, in which case the significance of your work will probably be overlooked. However, submitting your paper to prestigious journals such as *Science, Nature*, and *The New England Journal of Medicine* increases your chances of having the paper rejected. Indeed, these journals reject about 85% of the papers they receive, whereas the *Journal of Biological Chemistry*, a more specialized journal, rejects only about half of the papers that it receives. You can choose an appropriate journal for your paper by talking with colleagues and determining where the best papers in the discipline are published. Also examine *Journal Citation Reports*, a listing of a journal's impact based on the number of times scientists have cited the journal's papers.

Your choice of journals will help determine the impact of your paper. For example, Gregor Mendel's paper "Experiments with Plant Hybrids" describing his plant-breeding experiments was published in 1866. Although Mendel sent copies of the paper to 120 scientific societies and universities, it remained unknown and was ignored for almost 40 years. This was no accident: Mendel published the paper in an obscure journal— *Transactions of the Brünn Natural History Society*. Not until DeVries, Correns, and Tschermak all thought that they had independently reached the same conclusion in about 1900 did they discover Mendel's paper and realize Mendel's genius.[14]

Quality of Reproduction   If your paper includes photographs such as electron micrographs, choose a journal that will reproduce your micrographs well.

Circulation   Journals with a large circulation are usually read by more scientists than those with a smaller circulation. Journals report their circulation in their "Statement of Ownership" that's published each year, usually in the November or December issue. Here are the figures for 1990 for a few journals:

Scientific American	606,826
Natural History	546,345

[14]Gasking, E.B. 1959. "Why Was Mendel's Work Ignored?" *Journal of Historical Ideas* 20: 60–84.

Science Digest	230,000
New England Journal of Medicine	226,000
Science	153,000
Skeptical Inquirer	38,000
Nature	31,000
Science Teacher	25,000
Journal of Irreproducible Results	12,000
BioScience	12,000
Proceedings of the National Academy of Science	10,000
American Biology Teacher	9,413
Ecology	8,434
Biotechnology	8,000
Journal of Biological Chemistry	7,200
Biochemistry	6,011
Plant Physiology	5,185
The Journal of Cell Biology	5,123
Genetics	4,723
American Journal of Botany	4,233
Evolution	4,223
Ecological Monographs	3,800
Journal of Animal Ecology	3,500
Quarterly Review of Biology	2,725
Biological Bulletin	2,600
Lipids	2,600
Journal of Neurochemistry	2,500
The Botanical Review	1,749
Human Biology	1,654
Yeast	1,500
Molecular Biology and Evolution	1,000
Biophysics	1,000

Journal of the Alabama Academy of Science	800
Italian Journal of Biochemistry	600
Invertebrate Taxonomy	500
Actinomycetes	300
Newsweek	3,057,081

Speed of Publication   Most scientists are in a rush to publish their work. This rush to publish dates to the seventeenth century when, in an effort to force scientists to divulge their data, a secretary of the Royal Society of London invented the rule that priority goes to whoever publishes first, not to who discovers first. Scientists have been vying to publish first ever since. For example, Watson and Crick's seminal paper in the April 1953 issue of *Nature* was published "as is" just three weeks after it was submitted for publication (that paper is reprinted near the end of this chapter). Such haste to publish is also driven by journal editors wanting to publish "hot" papers. Indeed, the competition between journals such as *Science* and *Cell* is "like an outright war." Many worry that such rapid publication may prompt premature publication—that in their rush to publish, scientists and editors may cut corners and the review process may be compromised, leading to incorrect or incomplete work. Stanley Pons's and Martin Fleischmann's paper about cold fusion, published by the *Journal of Electroanalytical Chemistry* just four weeks after its receipt, is a precedent that no one wants to repeat.[15]

If you work in a fast-moving field or are in a rush to publish your work, consider the time required by the journal to publish your paper. Journals such as *American Journal of Botany* take more than a year to publish a paper, while others such as *Science* and *Cell* take an average of about 4.5 months. To hasten publication, some journals receive manuscripts on disk, thereby easing editing and typesetting. However, no major journals yet receive manuscripts via electronic mail.

Formatting, Typing, Packaging, and Mailing Your Manuscript

It's a damn poor mind that can think of only one
way to spell a word.
    —Andrew Jackson

Formatting

Papers published in scientific journals must be submitted in a format specified by the journal. This format is detailed in the journal's "Instruction to Authors."

[15]Roberts, Leslie. 1991. "The Rush to Publish." *Science* 251: 260–263.

For example, here are the instructions for how to prepare a manuscript for publication in *BioScience*:

Information for BioScience Contributors

How to Prepare a Manuscript

The BioScience Staff

The editors welcome manuscripts written for a broad audience of professional biologists and advanced students. *BioScience* publishes Articles, which are peer reviewed, summarizing recent advances in important areas of biological research. *BioScience* also publishes short opinion pieces in the Viewpoint section, longer opinion pieces and essays on policy issues important to biologists in the Roundtable section, essays on biology education in the Education section, and Letters pertaining to material in *BioScience*. In addition, The Biologist's Toolbox contains commentaries and descriptions and reviews of instruments and computer-ware relevant to the professional biologist. Finally, the special book issues of *BioScience* include an article on some aspect of book writing, publishing, or reading. To speed publication, contributions to these departments—Viewpoint, Roundtable, The Biologist's Toolbox, Education, and Books—are not formally peer reviewed but are accepted by the editor, generally after consultation with an outside advisor.

The editors reserve the right to edit all manuscripts for style and clarity. Contributions are accepted for review and publication on the condition that they are submitted solely to *BioScience* and will not be reprinted or translated without the publisher's permission. All authors must transfer certain copyrights to the publisher.

Articles

Articles should review significant scientific findings in an area of interest to a broad range of biologists. They should include background for biologists in disparate fields. The writing should be as free of jargon as possible. All articles, whether invited or independently submitted, undergo peer review for content and writing style. Articles must be no longer than 20–25 double-spaced typed pages, including all figures, tables, and references. No more than 50 references should be cited. Submit an original and four copies of all manuscripts along with a cover letter. List in your cover letter the name of colleagues who have reviewed your paper plus the names, addresses, and telephone numbers of four potential referees from outside your institution but within North America.

Viewpoint and Roundtable

The Viewpoint page and Roundtable essays may cover any topic of interest to biologists, from science policy to technical controversy. Viewpoints must not exceed two and a half double-spaced pages; Roundtable contributions may be up to 15 double-spaced pages. To meet format requirements, Viewpoint submissions may not have footnotes or a reference list. Roundtable contributions may include a few photographs, drawings, figures, or tables. They may also include up to 25 references.

Education and Books

The Education section includes observations and opinions on the teaching of biology, both to students and to the general public. Education manuscripts must not exceed 15 double-spaced pages and may cite no more than 25 references. The Book section of the *BioScience* special book issues contains an article on some aspect of book writing, publishing, or reading. The book reviews are generally solicited. If you are interested in writing a book review, contact the Book Review Editor.

Manuscript Preparation

Submission Submit an original and two copies of all manuscripts along with a cover letter to Editor, *BioScience*, 730 11th Street, NW, Washington, DC 20001-4584. Be sure to include your telephone number and FAX number, if available. Authors must obtain written permission to use in their articles any material copyrighted by another author or publisher. Include with your manuscript photocopies of letters granting permission; be sure credit to the source is complete.

Typing Use double-spacing throughout all text, tables, references, and figure captions. Type on one side only of 8½ × 11-inch white paper. Type all tables, figure captions, and footnotes on sheets separate from the text. Provide a separate title page with authors' names, titles, affiliations, and addresses; include a sentence or two of relevant biographical information and research interests.

Style Follow the *Council of Biology Editors Style Manual*, 5th edition (CBE 1983), for conventions in biology, except for references cited. For general style and spelling, consult *The Chicago Manual of Style*, 13th edition (Chicago 1982), and a dictionary such as *Webster's Third International Dictionary* (Gove 1968).

Abstract Manuscripts for Articles and Roundtable, Education, and Toolbox sections should include an informative abstract no longer than 50 words. The abstract is useful to the editors and reviewers. It is not published. Do not include a summary in the article.

Symbols, Acronyms, and Measurement Define all symbols and spell out all acronyms the first time they are used. All weights and measures must be in the metric system, SI units. In Articles and Toolbox contributions, abbreviations may be used for units of weight or measurement that describe data.

References Cited and Footnotes No more than 50 references should be cited in Articles and no more than 25 in Roundtable, Toolbox, and Education. Personal communications, unpublished data, and manuscripts in preparation should be cited in footnotes containing the data and source's name and affiliation.[1] Keep other footnotes to a minimum. Number text footnotes with consecutive superscript numerals. For footnotes in tables, use symbols (page 79, CBE 1983). In-text citations must take the form: (Author date). Multiple citations should be listed in alphabetical order: (Author date, Author date). Use the first author's name and "et al." for in-text citation of works with more than two authors or editors; list every author or editor in the "References cited" list. All works cited in the text must be listed alphabetically in References cited; works not cited in the text should not be listed. Follow the BIOSIS List of Serials for journal abbreviations; provide the full name of journals not listed there. Underline the titles of all books and journals. Refer to recent issues of *BioScience* for additional formatting; some examples:

- A journal article: Bryant, P. J., and P. Simpson. 1984. Intrinsic and extrinsic control of growth in developing organs. *Quart. Rev. Biol.* 59: 387–415.

- A book: Ling, G. N. 1984. *In Search of the Physical Basis of Life.* Plenum Press, New York.

- Chapter in book: Southwood, T. R. E. 1981. Bionomic strategies and population parameters. Pages 30–52 in R. M. May, ed. *Theoretical Ecology.* Sinauer Associates, Sunderland, MA.

- Technical report: Lassiter, R. R., and J. L. Cooley. 1983. Prediction of ecological effects of toxic chemicals, overall strategy and theoretical basis for the ecosystem model. EPA-600/3–83–084. National Technical Information Service PB 83–261–685, Springfield, VA.

- Meeting paper: O'Leary, D. S. 1982. Risks and benefits of cooperating with the media. Paper presented at the annual meeting of the American Association for the Advancement of Science, Washington, DC, 8 January 1982.

Word Processing

We encourage authors to enclose an IBM PC- or Wang OIS-compatible floppy disk, along with the typed or printed copies of the manuscript. We prefer text to be in ASCII, free of formatting and control characters. Text may also be sent via modem to Biotron (202-628-2427), MCI Mail (To: AIBS, 800/456-6245), BITNET (AIBS@GWUVM.BITNET) or INTERNET (AIBS@GWUVM.GWU, EDU).

Illustrations

Black-and-white Photographs, maps, line drawings, and graphs must be camera-ready, glossy black-and-white prints, photostats, or original art. On the reverse side, number and identify figures and indicate "top" of photographs. All photographs must be untrimmed and un-mounted, 4×5 to 8×10 inches in size, and as clear as possible; photomicrographs should have a scale bar. Line drawings and graphs should be done by professional artists or scientific illustrators. Lettering must be large enough to be legible after a 50% reduction.

Color Authors must pay the cost of printing color art within an article. Contact the Editor for details.

Cover Authors are encouraged to submit color transparencies for consideration as cover art. These photos should be related to the manuscript. Cover photographs must be in a vertical format, sharply focused, and colorful. They must have a light or dark area at the top for the logo.

References Cited

CBE Style Manual Committee (CBE). 1983. *Council of Biology Editors Style Manual: A Guide for Authors, Editors, and Publishers in the Biological Sciences.* Council of Biology Editors, Bethesda, MD.

The Chicago Manual of Style (Chicago). 1982. University of Chicago Press, Chicago.

Gove, P. B., ed. 1968. *Webster's Third New International Dictionary of the English Language Unabridged.* G. & E. Merriam Co., Springfield, MA.

CORRESPONDENCE: Direct all correspondence to *BioScience* Editor, American Institute of Biological Sciences, 730 11th Street, NW, Washington, DC 20001-4584. Tel: 202/628-1500; FAX: 202/628-1509.
[1]Footnote format: H. J. Smurd, 1989, personal communication, University or other affiliation, city, state.

The staff of *BioScience* warns authors, "Manuscripts submitted must follow the formats specified here." If you ignore the formats specified by the journal, the editor will return your paper to you.

Spelling

Long ago, people paid little attention to spelling. Indeed, Chaucer often changed the spelling of words to fit his rhymes. Today we have one correct way to spell each word (words having two spellings usually have one spelling that is preferred). We often accept different pronunciations, but not different spellings. Therefore, type your paper carefully, being sure to check the spelling of every word. Papers loaded with typos usually indicate that the writing and science are poor. For example, when people read *Noble laureate* for *Nobel laureate*, they usually question the writer's intelligence, education, and everything else that he or she has to say. This may not seem logical or fair, but it's the way it is. Moreover, typos can have important consequences: For example, spinach may have got its reputation as a dietary supplement because of a misplaced decimal point in which the iron content was reported as ten times higher than it was.[16] Other reports claim that typos have resulted in deaths.[17]

You'll probably not know how to spell every word that you use. However, don't be overly concerned; all writers have words that trouble them. For example, John Irving looked up *strictly* 14 times during one 5-year period; during the same time, he also looked up the word *ubiquitous* 20 times. Not only can he still not spell these words, he can barely remember what they mean. However, that's no excuse to misspell words. To avoid problems, always follow spelling's only rule:

IF YOU DON'T KNOW, LOOK IT UP.

Double-space your paper and type on only one side of each sheet. Use margins at least 3 cm wide and begin the Abstract on a new page. Place tables and figures at the end of the paper. Consult the journal's "Instructions to Authors" for such information as how many copies to submit.

Packaging and Mailing

Package your manuscript carefully. Editors are continually perplexed by authors who mail elaborate artwork and photographs in flimsy envelopes that are damaged or destroyed by the postal service. To help ensure that your paper arrives in good condition, mail the paper in a heavy-duty, padded envelope whose ends and seams are reinforced with tape. Use no staples: they can tear or scratch artwork. Keep a copy of your paper for your files.

Include a cover letter with the paper that states (1) that the work reported in the paper is original and has not been submitted elsewhere for publication; (2) the name and address of a person to contact about the paper; and (3) that all authors participated in the research. Don't bias the editor against your paper with a long, poorly written cover letter. For example, one author closed a cover letter with "I hope you will find this paper exceptable." The editor did.

What Happens Next

Within a week or two the journal's editor will send you a letter acknowledging receipt of your paper. The acknowledgment will probably also tell you when you can expect a decision about the fate of your paper. After you receive the letter of

---

[16]Hamblin, T.J. 1981. "Fake." *British Medical Journal* 283: 1671–1674.

[17]An article entitled "Mortal Consequences of a Typographical Error" describes how a German physician had read in a journal that cases of pruritus ani had been treated successfully with a 1% solution of percaine. When he gave this solution to a patient, the patient died. The physician was unaware that the journal's editor, in a subsequent issue of the same journal, had published a correction stating that the solution should have been one per thousand, not one per hundred. Nevertheless, a German court held the physician guilty of negligence on the grounds that, considering the novelty of the treatment, he should have followed the literature more closely and sought advice from competent authorities. (see "Consequences mortelle d'une erreur typographique." *Med dans le Monde* [Suppl. Semaine Med] 26: 537, 1950).

acknowledgment you'll hear nothing more about your manuscript for one or two months. During this time, the editor will send your paper to other experts (usually two) for their opinions about its publication in the journal. These reviewers recommend to the editor that the paper either be published or rejected. On the basis of these reviews and his or her opinion of the paper, the editor will then decide the fate of your paper.[18] This decision will usually be one of the following:

Publish the paper "as is." If this happens, congratulate yourself. Such a decision is rare: Fewer than 5% of papers are published "as is."

Publish the paper after it's revised according to suggestions of the editor and reviewers. About 45% of papers fit this category. In these cases, you must enclose with your revised manuscript a cover letter describing how you've met the reviewers' criticisms. If you've ignored any of the reviewers' suggestions, you must defend your actions. Such explanations are extremely important. Indeed, the fastest way to have your paper rejected is to return it without giving reasons for ignoring the reviewers' suggestions. Few editors will tolerate such arrogance and disregard for the criticism of your peers. However, if you make the suggested changes or present a good reason for not changing your paper, most editors will accept your paper for publication.

Reject the paper. About half of all papers submitted to top-quality journals are rejected, meaning that the editor has decided that your paper will not be published in the journal. If you get a rejection letter, read it carefully. Perhaps your paper was rejected because it describes research that is irrelevant to the scope of the journal. Although such decisions are pointless to challenge, remember that such a decision is not a criticism of your data or conclusions: You've merely sent your paper to the wrong journal. Other papers are rejected because of defects in the research, such as no data for a critical control experiment. In these cases, do the needed experiment and resubmit the paper. Finally, other papers are rejected because they represent poor science—poor experimental design

---

[18]A negative review will not necessarily cause a good editor to reject a paper from publication. Editors ignore reviews that include no evidence for the reviewers' recommendations. Indeed, reviews such as, "This paper shouldn't be published because it is not worthy of publication" tell editors and authors nothing. Good editors demand that reviewers document their recommendations with evidence.

The history of science is filled with examples of editors having ignored reviewers' recommendations to publish what became important papers. For example, shortly before Christmas 1671, Newton's reflecting telescope (the first ever built) was greeted with acclaim by the Royal Society. Soon thereafter, Newton told the Society that he would submit a paper that he considered "the oddest if not the most considerable detection which hath hitherto beene made in the operation of Nature." Robert Hooke, at that time England's leading authority on optics, was highly critical of Newton's paper, but his criticisms neither halted nor delayed publication. However, the matter did not end there. Hooke and Newton continued their bitter dispute for many more years. Scientists today often misjudge papers they review. For example, four of the six most-cited papers from *The Lancet* and *British Medical Journal* from 1955 to 1988 contained ideas that were initially rejected or disbelieved (see Dixon, Bernard. 1989. "Disbelief Greeted Classics in Top U.K. Medical Journals." *The Scientist* 17 April 1989).

and unfounded conclusions. In this case, reconsider the validity of your paper. You may need to revise your experiments.

Despite your best attempts, you'll probably receive a rejection letter someday. Do not despair: Most rejected papers are eventually published. For example:

> Hans Krebs's first paper on what came to be known as the Krebs cycle was published in *Enzymologia* two months after being rejected by *Nature*.[19]

> The *Journal of Clinical Investigation* rejects half of the papers it receives; 85% of the rejected papers are published elsewhere.[20]

A few months before your paper is published you'll receive page proofs of the paper. These are typeset pages of your paper for you to examine and correct. The editor or printer will give you detailed instructions about how to mark the page proofs and when they must be returned. Here are some suggestions for handling the page proofs:

Do not just read the page proofs. If you do, you'll have no way of catching mistakes such as "61" substituted for "16." Rather, *study* the page proofs carefully: They're the last you'll see of your paper before it's published. Any mistakes that you don't catch will appear in your paper when it is published. Such mistakes can damage your credibility and embarrass you. For example, consider this blooper that went undetected: "She was treated with Mycostatin oral suppositories."

Have someone read the manuscript aloud—word by word, line by line—while you check the accuracy of the proofs. After spending 8 months and $300,000 to gather data, don't sleep through a typesetter's change of your paper.

Read the corrected page proofs aloud to yourself. Speaking makes you think about what you wrote and helps highlight mistakes and bloopers such as these:

Patient has chest pains if she lays on her side for over a year.
Father died in his 90s of female trouble in his prostate and kidneys.
If he squeezes the back of his neck for 4 or 5 years it comes and goes.
The skin was somewhat pale but present.
The patient was mentally alert but forgetful.

Pay special attention to all numbers (dates, numbers, and formulas), scientific and proper names, punctuation, and figure placements.

Keep a copy of the corrected proofs and return one copy to the editor.

[19]Krebs, H.A. 1981. *Reminiscence and Reflections*. Oxford: Clarendon Press.
[20]Wilson, J.D. 1978. "Peer Review and Publication." *Journal of Clinical Investigation* 61: 1697–1701.

Near the time you get the page proofs you'll also receive a form for ordering reprints of your article. You'll probably want to order reprints, especially if your paper contains photographs that won't photocopy well.

A Final Word about Dealing with Editors

Editors of journals are the gatekeepers of scientific literature: They decide which papers will be published and which papers will be rejected. Although their decisions are usually guided by reviewers, it is the editor who has the final word on the fate of a paper. If you deal with editors courteously, you'll learn that they are reasonable scientists and are on your side: They *want* to publish high-quality papers in their journals. You'll get nowhere by conceitedly insisting that science can't advance unless your ideas are enshrined in print.

Other Kinds of Writing for Professional Audiences

Although you must learn to write research papers if you want a career in biological research, your writing tasks won't stop there. You'll probably also need to write review articles, abstracts, poster presentations, a thesis and dissertation, grants, and conference reports.

Review Articles

Reviews are papers that describe someone else's interpretation and evaluation of the primary literature. Reviews summarize, analyze, and critically evaluate literature already published. They resemble term papers in that they review the literature and, if they are effective, put it in a new perspective. Reviews do not report original research and are usually 10 to 50 pages long. The audience is more general than that for a research paper. Although review papers have no prescribed format, most concentrate on reviewing and discussing the literature. Useful reviews should try to answer important questions. They may not provide a definite answer but they should indicate directions for future research.

Review articles seldom contain discussions of materials and methods. They often are longer than research papers because they cover much material derived from many sources about a relatively broad subject. Reading review articles will immerse you in the biological literature, thereby showing you how biologists think. Reading these papers is also the best way to stay up-to-date on developments in fields other than your own. The following books and journals publish review articles for scientists and nonscientists:

> *Annual Reviews.* This series of review articles began in 1950 with publication of *Annual Review of Medicine.* There are now more than 30 versions of *Annual Reviews* (for example, *Annual Review of Plant Physiology*).

American Scientist. Review articles covering a variety of sciences.

Biological Reviews. Review articles on biological topics.

Physiological Reviews. Technical articles about physiology.

Quarterly Review of Biology. Excellent review papers about biology.

Scientific American. Readable articles about many scientific topics.

Many biologists write review articles in response to an invitation from a journal's editor. If you've not received such an invitation, ask the journal's editor if she or he is interested in publishing such a review. *Science Citation Index* codes reviews with an "R."

Theses and Dissertations

A thesis or dissertation is a statement, proposition, or position that a person advances and is prepared to defend. It describes original research and therefore is a type of research paper. Thesis requirements vary at different universities, and usually only the format of the thesis (for example, where to place page numbers and the sizes of margins) is prescribed. Theses usually take several months to write, so students should start writing them long before their anticipated date of graduation.

Many scientists question the validity of a traditional thesis. Since a thesis is a published report of original research, it should be written with the same rigor as a research paper. However, this is seldom the case. Many scientists think that theses differ from primary publications. Consequently, most theses are not subjected to the same criticism as a primary publication. This produces excessively long theses containing irrelevant and trivial data—50 pages of science crammed into a 250-page thesis. They're then filed away in the basement of the library, never to be opened again except perhaps by a bored janitor. J. Frank Dobie said it well: "The average Ph.D. thesis is nothing but a transference of bones from one graveyard to another."

Many theses are written poorly because most scientists do not understand writing. Indeed, many fear a "thin" thesis because they equate an article's length with its significance. Others defend a verbose thesis as critical to measuring characteristics such as mastery of the subject, investigation, and critical thinking. However, these characteristics are features of the student that are determined long before the thesis is written. A thesis should measure scientific preparation, not activity. Consequently, verbiage in a thesis indicates either laziness or an attempt to hide one's scientific shortcomings. A 250-page thesis containing trivial data, poor analyses, and a "comprehensive" literature review and bibliography shows only that a student has learned to read, write, and stuff his or her thesis with "filler" rather than to think or write effectively.

Remember this: There is no correlation between the length of a scientific paper and its significance. Similarly, including "everything" in a thesis merely indicates that the student hasn't learned to discriminate what's important from what's trivial and unnecessary. To quote an old saying, "Only fools collect facts; wise people use them."

Don't accept the argument that a thesis (or anything else that you write) must be long to be important. Get help from your major professor, and approach a thesis with the same rigor as you would a research paper. Meet the local requirements for the thesis, but remember that professional scientists publish research papers, not theses. Your success as a biologist will depend largely on your ability to write a research paper. Once you graduate, you'll never write a thesis again. Therefore, concentrate on learning to write high-quality papers for publication in high-quality, peer-reviewed journals. One journal article contributes much more to science than does a pile of poorly written theses filed in a dusty library.

Grants

Before the nineteenth century, science was a hobby of the rich. Members of the nobility and the literate bourgeoisie set aside rooms of their houses as scientific labs. People without the leisure afforded by wealth had to raise their own money to pursue science. For example,

> Johannes Kepler moonlighted as a court astrologer. His three laws of planetary motion, notably his deduction that planets move in elliptical orbits around the sun, vindicated Copernicus and served as the basis for Isaac Newton's research.

> Francis Bacon was a lawyer who became Attorney General and Lord Chancellor of England. His legal service ended when he was charged with taking a bribe. However, neither Bacon's public service nor his dismissal slowed his studies showing that nature could be confirmed and understood only by systematic experimentation.

> Antoine Lavoisier was a tax-farmer (a tax-collector empowered to make a personal profit) in France. He named oxygen and hydrogen, wrote the first textbook of modern chemistry, and is regarded as the father of modern chemistry. His profits as a tax-farmer supported his research but also were his downfall: Soon after the French Revolution, Lavoisier's career ended at the guillotine.

> Leonardo da Vinci's pathetic letter seeking a job of Lodovico Sforza emphasized his ability to contrive instruments of war as a means of supporting his scientific studies.

During the nineteenth century, scientists went out of their way to show the relevance of their work to the public as a means of obtaining money to support their work. For example, Thomas Edison sponsored an "electric breakfast" at which he cooked all the food with electricity. He used the profits of the demonstration to support his research. However, in the half century before World War I, the increasing professionalism of science burdened science with much jargon that isolated scientists from the public. Scientists began to speak only to their peers, causing much of the public to lose its understanding of science. Many scientists began to respect only "pure" research and viewed any popularization of science as vulgar. Soon thereafter, scientists' work began to be funded primarily by governmental grants, a trend that continues today. To get this governmental money, a scientist must write a grant proposal.

Grant proposals are written by scientists requesting money to support a research or teaching project. They're read and evaluated by a small audience— usually only a few people. Grant proposals have many forms, ranging from letters to long, involved proposals. The format and length of a proposal are determined by funding agencies to which the proposal is submitted, such as the National Science Foundation and the National Institutes of Health.

Obtaining grants is one of the most important and stressful aspects of being a research scientist. Although many books and seminars claim that following a simple set of rules will ensure getting a grant, scientists know better: No one set of instructions can guarantee that you'll get a grant. However, all scientists agree that you will increase your chances of getting a grant if you do the following things:

Propose an original, defined idea with a measurable conclusion. Be sure the experiments that you propose can be done within the time frame of the grant.

Understand the problem and the proposed work. Your proposal should demonstrate your knowledge of the field and the relevance of the proposed work.

Include a detailed plan and timetable for doing the work. Include preliminary data, if they are available.

Include a curriculum vitae documenting your credentials for doing the work.

Include a detailed budget for the work.

Many grant proposals are rejected because they are written poorly. Therefore, don't waste your time by throwing together a poorly written proposal that outlines trivial work or work that you can't do. Research funds are hard to get, even when a proposal is written well by a competent scientist. Poorly written proposals have no chance of succeeding. If you want money for your work, learn to write well.

Abstracts

"Reviewed" abstracts are abstracts written as part of a research paper. Scientists also write "nonreviewed" abstracts, which are summaries of research papers written to obtain a place on the program at a scientific meeting. Nonreviewed abstracts are published "as is" and without review, while reviewed abstracts are published only after the research paper is reviewed by experts. Nonreviewed abstracts are written like reviewed abstracts (see above).

Unlike conference reports, abstracts often contain original data and are therefore considered a type of primary literature. Most organizations and professional societies provide forms on which you must type the abstract.

Poster Presentations

Many biologists use posters to communicate their findings to colleagues at professional meetings. Indeed, poster sessions are the primary way that biologists exchange information at many meetings. Posters are displayed for several hours or days, thereby allowing biologists to exchange information rather than merely present data.

Posters are written summaries of papers and are organized like a research paper. They announce new results, contain few details, and should be visually interesting to help attract the target audience. Posters are usually allotted about 1.2 m $\times$ 2.5 m of space. Pattern your poster after the most effective posters that you see at a poster session at a professional meeting. Print the title in letters at least 3 cm high, the authors' names about 2 cm high, and the text in letters about 4 mm high. Use simple writing to highlight your most important data and conclusions.

Conference Proceedings

Each year there are about 10,000 scientific meetings, two thirds of which publish their proceedings. Biologists who attend these meetings are often required to write a conference report describing their presentations. These conference reports are usually 1,000 to 2,000 words long and seldom contain many details or historical perspectives. Furthermore, they are seldom reviewed before publication and often are published in "proceedings volumes" that are ignored by most other biologists. Conference reports are good places for speculation, tentative conclusions, and preliminary data. Since many of these data are published later in research papers, authors must beware of copyright problems (see "Copyright and Ownership").

Pranks, Hoaxes, and Faked Data

> Science does not select or mold specially honest people: it simply places them in a situation where cheating does not pay. . . . For all I know, scientists may lie to the IRS or to their spouses just as frequently or as infrequently as everybody else. — S.E. Luria

William Osler was a great biologist and writer; indeed, his celebrated textbook *Principles and Practice of Medicine* is a masterpiece of English literature and a monument to Osler as scientist, clinician, and teacher. However, Osler was addicted to pranks. He published at least 18 pranks under the pen name Egerton Y. Davis. Most of these pranks had a Rabelaisian bent, such as his description of "Peyronie's disease" or "*strabismus du penis*," in which "when erect it curved to one side in such a way as to form a semicircle, hopeless and useless for any practical purpose." Osler submitted this paper under the name of a prominent Philadelphia urologist, who responded by confirming the observation and signing it EYD Jr.* Interestingly, some of Osler's pranks were cited in other papers. For example, one writer cited the "Peyronie's disease" paper as being by "a medical man called Davis, not otherwise identified."†

Unfortunately, many biologists have published faked data. Among the most notorious stories of faked data is the story of the "midwife toad" reported by Austrian zoologist Paul Kammerer in the 1920s. Kammerer used midwife toads (*Alytes obstetricans*), which breed on dry land and therefore never develop pigmented "nuptial pads" that characterize related species that breed in water. When Kammerer bred the dryland toads in water, he claimed to have transmitted the ability to develop nuptial pads from one generation to another. This induced transmission of acquired characteristics provoked many evolutionary biologists. Other biologists were skeptical, and one visitor to Kammerer's lab saw where India ink was injected where pads were said to exist. Kammerer admitted fraud, but blamed the fraud on an unknown person. A few weeks later, Kammerer shot and killed himself.‡

Another famous case of faked data centered on James Shearer. In 1916, while British troops were dying in trenches of Europe, Shearer published a paper in the *British Medical Journal* describing a method he devised for depicting the path of gunshot wounds. British military officials quickly implemented Shearer's methods. However, it quickly became obvious that the data were faked, and the journal published a retraction. Shearer, who had been in His Majesty's Army, was sentenced to death by firing squad. The sentence was later commuted, and Shearer died of tuberculosis in prison a year later.§

Others have faked data with less serious consequences. For example, German zoologist Ernst Haeckel altered illustrations by labeling three copies of the same plate as human, dog, and rabbit to "prove" their developmental

---

*Nation, E.F. 1973. "William Osler on Penis Captivus and Other Urological Topics." *Urology* 2: 468–470; Teigen, P.M., and E.H. Bensley. 1981. "An Egerton Y. Davis Checklist." *Osler Library Newsletter* 38: 1–5.

†Taylor, F.K. 1979. "Penis Captivus—Did It Occur?" *British Medical Journal* 2: 977–978.

‡Koestler, A. 1972. *The Case of the Midwife Toad*. New York: Random House.

§Lock, S. 1988. "Misconduct in Medical Research: Does It Exist in Britain?" *British Medical Journal* 297: 1531.

similarities. He later defended himself against accusations of fakery by saying, "... hundreds of the best observers and biologists lie under the same charges."¶ Similarly, Pasteur may have been guilty of "creative reporting": Data recorded in his laboratory notebook differ from those in his papers about anthrax and rabies vaccines.|| Other examples of faked research include the following:**

A Yale researcher plagiarized others' work and falsified work in a dozen papers he coauthored with his department head.††

A Stanford professor published an article containing citations of papers not published.‡‡

A researcher in 1975 used a black felt-tipped marker to fake a successful skin-graft on a white mouse.§§

A more recent case of alleged fraud involved a paper published in the April 1986 issue of *Cell*. That paper came from Nobel laureate David Baltimore's laboratory at the Massachusetts Institute of Technology (MIT). The paper claimed that a foreign gene inserted into mice had influenced the way the mice produced antibodies, a finding that was described by many scientists as "surprising" and "important." When scientists couldn't repeat the results, the theory was quickly forgotten. However, the incident didn't die there. It became known as "the Baltimore affair" and today symbolizes the fallibility, arrogance, and deceit that sometimes accompany modern science.

What began as a laboratory dispute between Tufts University immunologist Thereza Imanishi-Kari and her then postdoctoral associate, Margot O'Toole, escalated into one of the most celebrated cases of alleged scientific fraud. The trouble started when O'Toole, unable to repeat results and puzzled by data purporting to support them, charged that a paper was based on data faked by Imanishi-Kari, one of six authors of the paper. Rather than refute these claims by repeating the experiments, the scientists, led by Baltimore, closed ranks. O'Toole suffered a fate common to whistle-blowers: She lost her job and was ostracized. Baltimore labeled her a malcontent. In a letter mailed to hundreds of his colleagues, Baltimore declared that the problem was "imagined" and warned that the incident "could cripple American science." Meanwhile Imanishi-Kari was hired at Tufts University. Baltimore also claimed that "the errors that have been identified in the *Cell* paper were inconsequential to the conclusions." Later investigations by two different panels at the National Institutes of Health (NIH) and by three Congressional hearings prompted many scientists to rally behind Baltimore and call the hearings a "witch hunt." However, the investigation continued and produced remarkable evidence. For example, the radiation-counter

¶Hamblin, T.J. 1981. "Fake." *British Medical Journal* 283: 1671–1674.
||Le Fanu, J. 1983. "Pasteur's Notes Tell a Different Story." *Medical News* 7: 9.
**In "Pathological Science" (Colloquium at the Knolls Research Lab, December 18, 1953), Irving Langmuir provides an amusing and tragic account of a series of "discoveries" (primarily by physicists) in which they found things that were never there. Interestingly, some of these "discoveries" generated hundreds of supporting studies.
††Broad, W.J. 1980. "Imbrogilo at Yale." *Science* 210: 38–41, 171–173.
‡‡"Stanford Denies Cover-up of Research Fraud." *New York Times* 23 August 1981, p. 31.
§§Broad, W., and N. Wade. 1982. *Betrayers of the Truth*. New York: Simon and Schuster.

tapes attached to Imanishi-Kari's notebooks were green. When the Secret Service studied more than 60 notebooks from other scientists in the same lab, they found no similar tapes dated later than January 1984. A more detailed study of tape color, type font, and ink produced what the Secret Service called a "full match" between Imanishi-Kari's tapes and those produced by Charles Maplethorpe, then a graduate student in her lab, between November 26, 1981, and April 20, 1982. This suggested to investigators that the tapes were not generated as part of the research for the *Cell* paper at all. The 121-page, minutely detailed draft report of the NIH's Office of Scientific Integrity:

> claimed that Imanishi-Kari committed "serious scientific misconduct" by "repeatedly present[ing] false and misleading information" to federal investigators.

> described O'Toole's actions as "heroic" and praised her "dedication to the belief that truth in science matters."

> claimed that Imanishi-Kari "fabricated" and "falsified" data.

> criticized Baltimore who, even after an earlier NIH panel found "significant errors of misstatement and omission" in the paper, called for "all scientists" to support Imanishi-Kari in the face of mounting evidence of fraud.

The statements of Baltimore's that the Office of Scientific Integrity found "most deeply troubling" were made to the NIH in 1990. These statements concerned data not included in the original paper, but which Imanishi-Kari gave to an NIH panel to support her published claims against O'Toole's charges. When the panel asked for the new data to be published as a correction to the original paper, Baltimore said, "In my mind you can make up anything you want in your notebooks, but you can't call it fraud if it wasn't published. Now, you managed to trick us into publishing—sort of tricked Thereza—into publishing a few numbers and now you're going to go back and see if you can produce those as fraud."

Throughout the investigation Baltimore had publicly battled all assaults on the integrity of his colleagues and of the paper itself. Baltimore's strategy was to try to divert attention from the original issue of the veracity of the data to one involving nonscientists policing science. Many scientists lined up to defend Baltimore without examining the facts of the case, apparently seduced by Baltimore's propaganda that the investigation could "cripple" science. Throughout the affair, Baltimore had tried to divert attention from the controversy by depicting the controversy as a political incident—a political attack on scientific freedom. Incredibly, Baltimore even claimed the NIH was somehow responsible for the fraud when he said, "If those data were not real, then she (Dr. Imanishi-Kari) was driven by the process of investigation into an unseemly act. . . ." However, in late March 1991, when drafts of the NIH report began to circulate, Baltimore suddenly tried to distance himself from the paper, saying that it raised "very serious questions." Moreover, on March 20, 1991 Baltimore finally asked that the *Cell* paper—the paper he had so adamantly defended—be retracted.¶¶ Despite his claims that he would not abandon a colleague in distress, he also then severed ties with Imanishi-Kari.

Although most scientists claim that such incidents are rare, other evidence suggests that fraud may be more common than is believed.‖‖ As of April 1991, NIH's Office of Scientific Integrity was investigating about 70 cases of suspected fraud. In a recent Acadia Institute survey of all graduate programs in the United States, 40 percent of deans reported that charges of misconduct or fraud had been brought to their attention. Most of the complaints were at major research universities. As this book went to press, another preliminary report from the NIH's office of Scientific Integrity had surfaced accusing Margit Hamosh, a pediatrics professor at Georgetown University, of lying and submitting "worthless and [apparently] fabricated or falsified" data in grant applications to the NIH and the Department of Agriculture. Interestingly, scientific fraud is a growth industry: Each month there are conferences and symposia on the topic. As you might guess, most retractions regarding faked data are ambiguous and loaded with doublespeak.

Scientists who publish faked data break the strong contract they have with their peers to honestly obtain, record, and publish their observations. Since everyone's work depends, in some way, on the work of others, scientists who consciously deceive other scientists by providing too few details to repeat their work, by reporting information that could lead others astray, and by publishing faked data should be regarded as a pariahs.

‖‖St. James-Robert, I. 1976. "Cheating in Science." *New Scientist* 72: 466–469; Altman, L., and L. Melcher. 1983. "Fraud in Science." *British Medical Journal* 286: 2003–2006.

¶¶Elmer-DeWitt, P. 1991. "Thin Skins and Fraud at MIT." *Time* 1 April 1991; Anonymous. 1991. "The Baltimore Affair: Ignoble." *The Economist* 30 March 1991; Hamilton, David P. 1991. "NIH Panel Finds Fraud in *Cell* Paper." *Science* 251: 1552–1554; Hamilton, David P. 1991. "Verdict in Sight in the 'Baltimore case.'" *Science* 251: 1168–1172.

James Watson and Francis Crick

Molecular Structure of Nucleic Acids

In the previous chapter you read part of *The Double Helix*, a best-selling book written by James Watson about his search for the structure of DNA. Here is the *Nature* paper that came from Watson's and Crick's research:

A Structure for Deoxyribose Nucleic Acid

We wish to suggest a structure for the salt of deoxyribose nucleic acid (D.N.A.). This structure has novel features which are of considerable biological interest.

A structure for nucleic acid has already been proposed by Pauling and Corey.[1] They kindly made their manuscript available to us in advance of publication. Their model consists of three intertwined chains, with the phosphates near the fibre axis, and the bases on the outside. In our opinion, this structure is unsatisfactory for two reasons: (1) We believe that the material which gives the X-ray diagrams is the salt, not the free acid. Without the acidic hydrogen atoms it is not clear what forces would hold the structure together, especially as the negatively charged phosphates near the axis will repeal each other. (2) Some of the van der Waals distances appear to be too small.

Another three-chain structure has also been suggested by Fraser (in the press). In his model the phosphates are on the outside and the bases on the inside, linked together by hydrogen bonds. This structure as described is rather ill-defined, and for this reason we shall not comment on it.

We wish to put forward a radically different structure for the salt of deoxyribose nucleic acid. This structure has two helical chains each coiled round the same axis (see diagram). We have made the usual chemical assumptions, namely, that each chain consists of phosphate

[1]Pauling, L., and Corey, R. B., *Nature*, 171, 346 (1953); *Proc. U.S. Nat. Acad. Sci.*, 39, 84 (1953).

This figure is purely diagrammatic. The two ribbons symbolize the two phosphate-sugar chains, and the horizontal rods the pairs of bases holding the chains together. The vertical line marks the fibre axis.

diester groups joining β-D-deoxyribofuranose residues with $3',5'$ linkages. The two chains (but not their bases) are related by a dyad perpendicular to the fibre axis. Both chains follow right-handed helices, but owing to the dyad the sequences of the atoms in the two chains run in opposite directions. Each chain loosely resembles Furberg's[2] model No. 1; that is, the bases are on the inside of the helix and the phosphates on the outside. The configuration of the sugar and the atoms near it is close to Furberg's "standard configuration," the sugar being roughly perpendicular to the attached base. There is a residue on each chain every 3–4 A. in the z-direction. We have assumed an angle of 36° between adjacent residues in the same chain, so that the structure repeats after 10 residues on each chain, that is, after 34 A. The distance of a phosphorus atom from the fibre axis is 10 A. As the phosphates are on the outside, cations have easy access to them.

The structure is an open one, and its water content is rather high. At lower water contents we would expect the bases to tilt so that the structure could become more compact.

The novel feature of the structure is the manner in which the two chains are held together by the purine and pyrimidine bases. The planes of the bases are perpendicular to the fibre axis. They are joined together

[2]Furberg, S., *Acta Chem. Scand.*, 6, 634 (1952).

in pairs, a single base from one chain being hydrogen-bonded to a single base from the other chain, so that the two lie side by side with identical z-co-ordinates. One of the pair must be a purine and the other a pyrimidine for bonding to occur. The hydrogen bonds are made as follows: purine position 1 to pyrimidine position 1; purine position 6 to pyrimidine position 6.

If it is assumed that the bases only occur in the structure in the most plausible tautometric forms (that is, with the keto rather than the enol configurations) it is found that only specific pairs of bases can bond together. These pairs are: adenine (purine) with thymine (pyrimidine), and guanine (purine) and cytosine (pyrimidine).

In other words, if an adenine forms one member of a pair, on either chain, then on these assumptions the other member must be thymine; similarly for guanine and cytosine. The sequence of bases on a single chain does not appear to be restricted in any way. However, if only specific pairs of bases can be formed, it follows that if the sequence of bases on one chain is given, then the sequence on the other chain is automatically determined.

It has been found experimentally[3,4] that the ratio of the amounts of adenine to thymine, and the ratio of guanine to cytosine, are always very close to unity for deoxyribose nucleic acid.

It is probably impossible to build this structure with a ribose sugar in place of the deoxyribose, as the extra oxygen atom would make too close a van der Waals contact.

The previously published X-ray data[5,6] on deoxyribose nucleic acid are insufficient for a rigorous test of our structure. So far as we can tell, it is roughly compatible with the experimental data, but it must be regarded as unproved until it has been checked against more exact results. Some of these are given in the following communications. We were not aware of the details of the results presented there when we devised our structure, which rests mainly though not entirely on published experimental data and stereochemical arguments.

It has not escaped our notice that the specific pairing we have postulated immediately suggests a possible copying mechanism for the genetic material.

Full details of the structure, including the conditions assumed in building it, together with a set of co-ordinates for the atoms, will be published elsewhere.

We are much indebted to Dr. Jerry Donohue for constant advice and criticism, especially on interatomic distances. We have also been

[3]Chargaff, E., for references see Zamenhof, S., Brawerman, G., and Chargaff, E., *Biochim. et Biophys. Acta*, 9, 402 (1952).

[4]Wyatt, G. R., *J. Gen. Physiol.*, 36, 201 (1952).

[5]Ashbury, W. T., *Symp. Soc. Exp. Biol.* 1, Nucleic Acid, 66 (Camb. Univ. Press, 1947).

[6]Wilkins, M. H. F., and Randall, J.T., *Biochim. et Biophys. Acta*, 10, 192 (1953).

stimulated by a knowledge of the general nature of the unpublished experimental results and ideas of Dr. M. H. F. Wilkins, Dr. R. E. Franklin and their co-workers at King's College, London. One of us (J. D. W.) has been aided by a fellowship from the National Foundation for Infantile Paralysis.

Compare the writing style in this article with that of *The Double Helix*. How do the writing styles differ? For example, how do the terminology, sentence length, and title differ in the two articles? What does the *Nature* article conceal? How does each style of writing achieve its purpose?

Charles Darwin

The Voyage of the *Beagle*

Charles Darwin (1809–1882) was part of a science-based family: His father was a rich physician, and his grandfather, Erasmus Darwin, was a physician–poet and amateur naturalist. When he was young, Darwin abandoned his study of medicine (and, in doing so, greatly disappointed his family) because he was horrified by operations on children without anesthesia. However, Darwin did remain interested in natural history. After graduating from Cambridge at the age of 22, he set sail aboard the H.M.S. *Beagle* because he thought it would be an opportunity for "collecting, observing and noting anything worthy to be noted in natural history."[21]

The cruise was planned to last only two years but eventually stretched to almost five. While sailing along the coast of South America, Darwin saw small variations in species. The most striking of these variations was among finches scattered among the Galapagos Islands. These observations made him question the immutability of species and were the basis for his ideas about evolution. Twenty years after returning to England, he published *On the Origin of Species* (1859), starting a furor over whether evolution applied to humans. Darwin stood his ground and, in 1881, published *The Descent of Man.*[22]

Here are a few paragraphs from *The Voyage of the Beagle*. The first paragraph tells why the voyage was formed:

[21]Contrary to myth, Darwin was not the naturalist aboard the H.M.S. *Beagle* when it set sail. The ship's surgeon, Robert McKormick, originally held the official position of naturalist (Darwin disliked McKormick's type of science). When McKormick was sent home to England in April of 1832, Darwin became the official naturalist aboard the ship. For more about this interesting story, see Chapter 2 of *Ever Since Darwin: Reflections in Natural History* by Stephen Jay Gould (W.W. Norton & Co., 1977).

[22]Although parts of Darwin's writing are criticized as stodgy Victorian prose—obscure, convoluted, and ambiguous—his *On The Origin of Species* is a masterpiece of organization, technical description, reporting, induction, and documentation. Darwin's logic forced scientists to accept his theory of natural selection despite all of the opposition that dogma, ignorance, and prejudice could muster. For a more lively and direct writing about the subject for a general audience, read the works of Thomas Huxley, a friend and advocate of Darwin's.

After having been twice driven back by heavy south-western gales, Her Majesty's ship *Beagle*, a ten-gun brig, under the command of Captain Fitz Roy, R. N., sailed from Devonport on the 27th of December, 1831. The object of the expedition was to complete the survey of Patagonia and Tierra del Fuego, commenced under Captain King in 1826 to 1830 — to survey the shores of Chile, Peru, and some islands in the Pacific — and to carry a chain of chronometrical measurements round the world. On the 6th of January we reached Teneriffe, but were prevented landing, by fears of our bringing the cholera: the next morning we saw the sun rise behind the rugged outline of the Grand Canary island, and suddenly illumine the Peak of Teneriffe, whilst the lower parts were veiled in fleecy clouds. This was the first of many delightful days never to be forgotten. On the 16th of January, 1832, we anchored at Porto Praya, in St. Jago, the chief island of the Cape de Verd archipelago.

The following paragraph tells us that Darwin is a keen observer:

The neighborhood of Porto Praya, viewed from the sea, wears a desolate aspect. The volcanic fires of a past age, and the scorching heat of a tropical sun, have in most places rendered the soil unfit for vegetation. The country rises in successive steps of table-land, interspersed with some truncate conical hills, and the horizon is bounded by an irregular chain of more lofty mountains. The scene, as beheld through the hazy atmosphere of this climate, is one of great interest; if, indeed, a person, fresh from the sea, and who has just walked, for the first time, in a grove of cocoa-nut trees, can be a judge of anything but his own happiness. The island would generally be considered as very uninteresting: but to any one accustomed only to an English landscape, the novel aspect of an utterly sterile land possesses a grandeur which any more vegetation might spoil. A single green leaf can scarcely be discovered over wide tracts of the lava plains; yet flocks of goats, together with a few cows, contrive to exist.

Almost 4 years into the voyage, the H.M.S. *Beagle* was in the Galapagos archipelago, the expedition's landfall. Darwin describes a lizard that even he seems to regard as an evolutionary oddball:

The Amblyrhynchus, a remarkable genus of lizards, is confined to this archipelago. There are two species, resembling each other in general form, one being terrestrial and the other aquatic. The latter species (A. cristatus) was first characterized by Mr. Bell, who well foresaw from its short, broad head, and strong claws of equal length, that its habits of life would turn out very peculiar, and different from those of its nearest ally, the Iguana. It is extremely common on all the islands throughout the group, and lives exclusively on the rocky sea-beaches, being never found, at least I never saw one, even ten yards in-shore. It is a hideous-looking creature, of a dirty black color, stupid, and sluggish in its movements. The usual length of a full-grown one is about a yard, but there are some even four feet long; a large one weighed twenty pounds.

Their tails are flattened sideways, and all four feet partially webbed. They are occasionally seen some hundred yards from shore, swimming about. . . . This lizard swims with perfect ease and quickness, by a serpentine movement of its body and flattened tail—the legs being motionless and closely collapsed on its sides. A seaman on board sank one, with a heavy weight attached to it, thinking thus to kill it directly; but when, an hour afterwards, he drew up the line, it was quite active. Their limbs and strong claws are admirably adapted for crawling over the rugged and fissured masses of lava, which everywhere form the coast. In such situations, a group of six or seven of these hideous reptiles may oftentimes be seen on the black rocks, a few feet above the surf, basking in the sun with outstretched legs.

I opened the stomachs of several and found them distended with minced sea-weed (Ulvae), which grows in thin foliaceous expansions of bright green or a dull red color. I do not recollect having observed this sea-weed in any quantity on the tidal rocks; and I have reason to believe it grows at the bottom of the sea, at some little distance from the coast. If such be the case, the object of these animals occasionally going out to sea is explained. The stomach contained nothing but the seaweed. Mr. Bynoe, however, found a piece of a crab in one; but this might have got in accidentally, in the same manner as I have seen a caterpillar, in the midst of some lichen, in the paunch of a tortoise.

The intestines were large, as in other herbivorous animals. The nature of this lizard's food, as well as the structure of its tail and feet, and the fact of its having been seen voluntarily swimming out at sea, absolutely prove its aquatic habits; yet there is in this respect one strange anomaly, namely, that when frightened it will not enter the water. Hence it is easy to drive these lizards down to any little point overhanging the sea, where they will sooner allow a person to catch hold of their tails than jump into the water. They do not seem to have any notion of biting; but when much frightened they squirt a drop of fluid from each nostril. I threw one several times as far as I could, into a deep pool left by the retiring tide; but it invariably returned in a direct line to the spot where I stood. It swam near the bottom, with a very graceful and rapid movement, and occasionally aided itself over the uneven ground with its feet. As soon as it arrived near the edge, but still being under water, it tried to conceal itself in the tufts of sea-weed, or it entered some crevice. As soon as it thought the danger was past, it crawled out on the dry rocks and shuffled away as quickly as it could.

I several times caught this same lizard, by driving it down to a point, and though possessed of such perfect powers of diving and swimming, nothing would induce it to enter the water; and as often as I threw it in, it returned in the manner above described. Perhaps this singular piece of apparent stupidity may be accounted for by the circumstance that this reptile has no enemy whatever on shore, whereas at sea it must often fall a prey to the numerous sharks. Hence, probably, urged by a fixed and hereditary instinct that the shore is its place of safety, whatever the emergency may be, it there takes refuge. . . .

I asked several of the inhabitants if they knew where it laid its eggs; they said that they knew nothing of its propagation, although well acquainted with the eggs of the land kind—a fact, considering how very common this lizard is, not a little extraordinary.

Darwin's writing is useful because it contains much information. Is it also personal? Why do you say so? Why are keen observation and detailed notes important to a scientist? To other professionals?

Exercises

1. Examine a paper published in *Ecology, Evolution, Cell, Journal of Animal Physiology*, or *Plant Physiology*. What hypothesis were the biologists testing? What were their assumptions? Do you agree with their conclusions? How would you rewrite the paper?

2. Read the "Letters" section of an issue of *Science*. What is the purpose of each letter? To provide information? To persuade?

3. Go to a library and examine a thesis, a research paper, a review article, a book, and an article written for a nonscientist. How does the writing differ? How is it similar?

4. Reread the Materials and Methods section about digestion of DNA by endonucleases (see page 246). What information did the authors omit from the paragraph? Why? Edit the paragraph to improve its clarity and readability.

5. Label the function of each sentence in these Introductions taken from research papers published in the *Journal of Animal Ecology and Journal of Protozoology*, respectively. Edit each Introduction to improve its clarity and readability.

Analysis methods for estimating age-specific survival rates from the ringing of young birds have generally been based on a single underlying model, or special case of this model. Our objective is to review this model that we will call the life table model, its assumptions, approaches to estimation of the model's unknown parameters and problems in making inference concerning age-specific survival rates. Our work was motivated, to some extent, by the recent paper by Lakhani & Newton (1983).

Species of *Balantidium* have been reported from at least six species of freshwater fishes (1, 2, 6–8) and one salmonid (*Salvelinus fontinalis* Mitch.) also from a freshwater environment (4). In each, the *Balantidium* is an intestinal symbiont and might be a parasite (7, 8). Reported here is the occurrence of *Balantidium prionurium* n. sp. from the intestinal lumen of the herbivorous saltwater surgeonfish, *Prionurus punctatus*.

6. Improve the clarity and readability of this Results section from a paper published in *Comparative Biochemistry and Physiology*:

In a survey of the ability of various cell-free and broken cell preparations of pseudoplexaurid-derived zooxanthellae to form squalene, first indications of success came on trial of a breaking medium with maleate as a buffer and the use of a Hughes press for cell breakage, see Table 1. The source of zooxanthellae in this case was *Pseudoplexaura wagenaari*. The other pseudoplexaurids, *P. flagellosa* and *P. porosa* also provided active preparations when their zooxanthellae were similarly treated. Since *P. porosa* typically yields more zooxanthellae and is common and easily identified in the field, its zooxanthellae were routinely employed in subsequent studies. Regardless of the source of zooxanthellae, maximally active preparations required both NaCl and KCl in the breaking medium. Breaking media employing buffers other than maleate gave less active or inactive preparations. This is not solely a pH effect since both medium III (MES buffer) and medium IV (phosphate) are at the pH of the maleate medium. Reduced activity is in part the consequence of inadequate breakage or of an inhibition of the squalene synthetic activity since transfer from breaking medium V (tris) to a maleate incubation medium results in moderate activity. The activity survived freezing whether frozen as intact zooxanthellae or as the broken maleate preparation, but with substantial loss over the period of a week. Consequently, only fresh preparations were employed.

7. Reread the abstract on page 241 of this chapter. What key words would you use to characterize this article? Here are the key words chosen by the authors: abscisic acid, corn, fluridone, viviparous, *Zea mays*. Do you agree with these choices? Why or why not?

8. Photocopy a section of a thesis written by a former student in the biology department. Edit the writing. What did you learn as you read and edited the writing?

9. Write an abstract for an article of your choice in *Scientific American*.

10. Discuss, support, or refute the ideas in these quotations:

The laboratory . . . I cannot write this word without emotion. For me, as for all my colleagues, it evokes so many memories, so many events, emotions, hopes. The laboratory is not only the setting of our material life, the place where a great part of our existence has been spent, but is above all, the nucleus of our intellectual life and even, at times of our sentimental life. — Pierre Lecomet du Nouy

The total energy of the universe is constant and the total entropy is continually increasing. — Isaac Newton

In the year 1657 I discovered very small living creatures in rain water. — Antonie van Leeuwenhoek

You can observe a lot just by watching. — Yogi Berra

Chance favors the trained mind. — Louis Pasteur

Every scientific fulfillment raises new questions; it asks to be surpassed and outdated. — Max Weber

Most institutions demand unqualified faith; but the institution of science makes skepticism a virtue. — Robert K. Merton

Science is what you know; philosophy is what you don't know. — Bertrand Russell

There are in fact two things, science and opinion; the former begets knowledge, the latter ignorance. — Hippocrates

True science teaches, above all, to doubt and be ignorant. — Miguel de Unamuno

Would you like to learn science? Begin by learning your own language. — Abbé de Condilla

Writing is an exploration. You start from nothing and learn as you go. — E. L. Doctorow

UNIT THREE

PRESENTING

INFORMATION

A picture shows me at a glance what it takes
dozens of pages of a book to expound.
  —Turgenev

Speech is the representation of the mind, and
writing is the representation of speech.
  —Aristotle

First learn the meaning of what you say, and then
speak.
  —Epictetus

It's difficult to think of any skill that is more
essential for career success than being able to write
well. Most often it's the written word that conveys
the personality, the ideas, and beliefs of the
individual. It's worth all the effort necessary to do
it well.
  —Burnell R. Roberts, Chairman and Chief
    Executive Officer, Mead Corporation

A tabular presentation of data is often the heart of, better, the brain, of a scientific paper.
   —Peter Morgan

Anything is better than not to write clearly. There is nothing to be said against lucidity, and against simplicity only the possibility of dryness. This is a risk well worth taking when you reflect how much better it is to be bald than to wear a curly wig.
   —Somerset Maugham

Too much talk will include errors.
   —Burmese proverb

There are no dull subjects. There are only dull writers.
   —H.L. Mencken

Everyone spoke of an information overload, but what there was in fact was a non-information overload.
   —Richard Saul Wurman

Every picture tells a story, don't it?
   —Rod Stewart

CHAPTER EIGHT

Numbers, Tables, and Figures

For he who knows not mathematics cannot know
any other science; what is more, he cannot discover
his own ignorance, or find its proper remedy.
    —Roger Bacon

Numerical precision is the very soul of science.
    —Sir D'Arcy Thompson

All science as it grows toward perfection becomes
mathematical in its ideas.
    —Alfred North Whitehead

Mathematics is both the door and the key to the
sciences.
    —Roger Bacon

Statistics are no substitute for judgment.
    —Henry Clay

Numbers

Science removes the fuzz and mystery from the world by using experiments and precise language. The most precise language is math. For example, there was no doubt that Galileo reported four—not three, not five—moons around Jupiter.

Biologists rely on numbers to describe many aspects of life, such as sizes of animals, heights of seedlings, and numbers of base-pairs in genes. Similarly, words such as *larger* and *smaller* are merely verbal expressions of numbers. Consequently, your ability to understand and write about numbers will affect your success as a biologist. To show this, consider the work of Gregor Mendel. Mendel's experiments with pea plants weren't original; several other biologists had done the same crosses and noted the same kinds of offspring seen by Mendel. Mendel's work differed from that of others in only one regard: Mendel counted his results. This allowed Mendel to determine the ratios that helped him explain some kinds of inheritance.

Biologists must know how to write effectively about numbers. This involves knowing what measurements to make, how to manipulate numbers statistically, and how to write about numbers.

Measurements in Science

Improved technology has greatly improved the accuracy of many of our measurements. For example, consider the speed of light, the one measure that stands as the most absolute in all of science. In late 1972, the accuracy of the measurement of that speed was increased 100-fold by Ken Evenson and his colleagues at the National Bureau of Standards when they estimated the speed of light to an accuracy of 0.5 meter per second. Such accuracy seems remarkable, but we still do not know exactly how fast light travels. All we have is a more accurate estimate. Nevertheless, such improved accuracy can have many effects. For example, before 1983 we determined the length of a meter independently, and the speed of light was specified by the length of a meter. Now, because the speed of light is considered constant, it defines the length of a meter: A meter is defined as the distance light travels in $1/299{,}792{,}458$ of a second. Any straight-edge that long is a meter long. Not surprisingly, we typically use indirect and inexact methods to make a measuring stick one meter long.

Another constant often used by biologists is Avogadro's number, which is the number of molecules in one mole (gram-molecular weight) of a substance. Avogadro's number is given as 6.02486×10^{23} molecules, but there is a 0.0027% error in this estimate. Although that error is relatively small, it comes to about $16{,}000{,}000{,}000{,}000{,}000{,}000$ molecules. Stated another way, Avogadro's number is accurate to about 16 quintillion molecules.

Do not be alarmed by this lack of complete accuracy and certainty in measurements. It simply means that many of our measurements can be improved.

The International System of Units

Biologists report measurements with a type of metric system called The International System of Units. This system is abbreviated SI and is the standard system of measurement used by scientists around the world. The system has several advantages, including that each quantity in SI has only one associated unit.

Measurements used most often by biologists include distance, mass, amount of substance, time, temperature, electric current, and luminous intensity. Prefixes to convert these so-called "base units" to other units—for example, to change meters to kilometers—are listed in Appendix 6.

Distance

The SI unit to measure distance is the **meter** (m). Area is measured as square meters (m²), and volume is measured as cubic meters (m³). Although the hectare and liter are used often by biologists, they are not SI units.

Here are equations to convert some English units to SI units:

1 m = 39.4 inches = 1.1 yd

1 km = 1000 m = 0.62 mi

1 ha = 10,000 m² = 2.47 acres

1 in = 2.54 cm

1 ft = 30.5 cm

1 yd = 0.91 m

1 mi = 1.61 km

1 L = 2.1 pt = 1.06 quart = 0.26 gal = 1,000 ml = 0.001 m³

1 ml = 0.03 fl oz

Mass

The SI unit to measure mass is the **kilogram** (kg). The kilogram is the only SI unit containing a prefix (kilo). Units derived from kilogram measure force (newton, N), pressure (pascal, Pa), work and energy (joule, J), and power (watt, W). The calorie is not an SI unit because it can't be defined without using an experimentally derived factor.

Remember that mass and weight are not equivalent. Mass measures an object's potential to interact with gravity and cannot be measured directly. Weight is force exerted by gravity on an object. An object that is weightless in outer space has the same mass as it has on earth. For more on misused words such as these, see Appendix 2.

Water Boiling at Zero Degrees Celsius?

The Celsius temperature scale was developed in 1742 by Anders Celsius, a Swedish astronomer. Celsius originally set the boiling point of water at 0°C and the freezing point of water at 100°C. J.P. Christine later revised the scale to its present-day form, with water boiling at 100°C and freezing at 0°C. Therefore, °C is appropriate in any case, but "degrees Christine" is technically more accurate than "degrees Celsius."

Amount of a Substance

The SI unit to measure the amount of a substance is the **mole** (mol). This mass is numerically equal to the molecular weight of a substance. The elementary entities (for example, atoms or ions) must be specified in the measurement. Concentration is expressed as moles per cubic meter ($mol \cdot m^{-3}$).

Time

The SI unit to measure time is the **second** (s). Units derived from the second are used to measure frequency (hertz, Hz; $cycles \cdot s^{-1}$), speed ($m \cdot s^{-1}$), and acceleration ($m \cdot s^{-2}$).

Temperature

The SI unit to measure temperature is the **kelvin** (K). Zero kelvin (0 K) is absolute zero, the hypothetical temperature characterized by the absence of heat and molecular motion. There are no prefixes for kelvin. Since temperature measures the intensity of molecular motion, all Kelvin temperatures exceed zero. A kelvin unit is the same as a degree Celsius, and to get from Celsius to kelvin, add 273 (K = °C + 273). Thus, water freezes at 273 K, room temperature is about 293 K, and body temperature is about 310 K. Note that kelvin measurements lack degree signs (°).

Most scientists use the Celsius scale to measure temperature. This scale uses 100 units to separate the temperatures at which water freezes and boils. Convert Fahrenheit temperatures to Celsius temperatures with this formula:

$$5(°F) = 9(°C) + 160$$

Electric Current

The SI unit to measure electric current is the **ampere** (A). Units derived from the ampere measure force or potential (volt, V) and electrical resistance (ohm).

Luminous Intensity

The SI unit to measure luminous intensity is the **candela** (cd). Units derived from the candela measure flux (lumen, lm). An incandescent light emits 10 to 20 lm for each watt of energy consumed, while a fluorescent light emits about 60 or more lumens for each watt.

Writing About SI Measurements

Follow these conventions when working with SI units and measurements:

SI uses symbols, not abbreviations. Use a period after a symbol only at the end of a sentence. SI symbols are always singular.

1.5 kg, not 1.5 kg.

10 km, not 10 kms

Symbols for prefixes are lowercase letters except for symbols denoting a million or more.

M	(mega; 10^6)	P	(peta; 10^{15})	G	(giga; 10^9)
E	(exa; 10^{18})	T	(tera; 10^{12})		

Write symbols immediately after their prefix.

kg, not k g

State units completely. Do not use a prefix alone and do not use compound prefixes.

km, not kilo

nm, not mmm

Insert a space between the number and the unit. However, do not space between the symbol and its prefix.

10 km, not 10km

Write symbols in lowercase letters except for unit names derived from a person's name.

A	(ampere)	J	(joule)	Hz	(hertz)
K	(kelvin)	W	(watt)	V	(volt)
Pa	(pascal)	N	(newton)		

Use a centered dot for a compound unit formed by multiplying two or more units.

$kg \cdot sec^{-1}$, not $kg \times sec^{-1}$

Express measurements with units requiring the least number of decimals.

1 km, not 1,000 m or 10^3 m

Do not mix symbols or units.

6.2 m, not 6 m 200 mm

Statistics

There are three kinds of lies — lies, damned lies and statistics.
     —Mark Twain

Statistics is a tool used to answer general questions on the basis of only a limited amount of specific information. Since statistics involves basing a judgment on a sample of a population (rather than the entire population), it helps us make a decision with incomplete knowledge of the population. Although this sounds unscientific, we do it all the time. For example, we diagnose disease with only a drop of blood, and we judge watermelons with only a few thumps on their rinds. Statistical studies of population samples are necessary when it is impossible or unrealistic to test the entire population.

Elementary principles of statistics are best described with examples. Suppose you have two samples of corn plants, each containing three plants that you grew in different kinds of light. The heights of the plants in the two samples are as follows:

SAMPLE #1: 6.0 cm, 6.0 cm, 6.0 cm

SAMPLE #2: 9.0 cm, 7.0 cm, 2.0 cm

In Sample #1, there is no variation: The heights are equal. Although plants in Sample #2 have the same mean height (6.0 cm), Sample #2 varies more than does Sample #1. This illustrates an important point: Simply listing the mean of a sample doesn't give readers the whole story. What, then, do you need to report?

Almost all biological data are variable. This variability results from biological variation, imprecise measurements, and the imperfections of machines we use to make our measurements. For example, if you remeasured the plants in your samples, you might get data like these:

SAMPLE #1: 5.9 cm, 5.9 cm, 6.1 cm

SAMPLE #2: 9.2 cm, 7.1 cm, 2.1 cm

These measurements have produced different means. Is this difference due to biological differences between the samples, a small sample size, or the imprecision of your measurements? The silo of seeds from which you picked your six experimental plants may have contained thousands of other seeds: How do you know that you didn't pick the six that are most different from the "average" seed in the silo? And how do you know that the seeds in your silo are like the corn seeds in other silos? One way to test this would be to increase your sample size—to repeat the experiment with, say 1,000 seedlings. Better yet, measure 100,000 seedlings. Better yet, measure 1,000,000,000 seedlings. Better yet, . . .

The only way to know "the truth" and be certain of your conclusions would be to measure all the corn seedlings in the world. Since this is impossible, we must measure seedlings that *represent* all the corn seedlings in the world. Such representative samples only estimate the truth: They only tell us how convincing or wacky our results are. Similarly, statistics tell us how representative our samples are.

Most numerical data must be subjected to appropriate statistical tests to determine if differences are substantial enough to be convincing. Indeed, many journals require that you use statistical tests to verify your results. Here are things to remember when using statistical tests to compare numerical data:

Define a hypothesis, which is the issue that you'll test. This hypothesis is termed the null hypothesis and is designated H_0. The null hypothesis assumes nothing unusual has happened. In our experiment, the null hypothesis states that the different kinds of light do not affect the height of corn plants. Although your data may be convincing, they cannot prove a hypothesis. Rather, they can only discredit or support the hypothesis.

Statistics reduces a population to a few characteristic numbers. Statistics are not data, and you lose information as you reduce your population to its characteristic numbers. The farther you reduce your data, the farther you remove yourself from the data. Therefore, keep the statistics as simple as you can.

Decide on a reasonable degree of risk (usually 5%) of incorrectly rejecting the null hypothesis. No matter what statistical test you use, there is always a chance that you will make the wrong decision. All that a statistical test will tell you is the *chance* of making that wrong decision. You cannot prove a hypothesis with statistics.

Common statistical tests include Student's *t*, Chi-square, and analysis of variance (ANOVA). The type of test that you'll use will depend on the amount and type of your data and the nature of your null hypothesis. Each test manipulates data in standard formulas and produces a test number. Test numbers near zero suggest that your data are consistent with the null hypothesis, while test numbers far from zero suggest that the null hypothesis is incorrect.

Look up the test number in a standard table to see if the number is within the expected range of values. If your test number falls within this range, your data support the null hypothesis. If your test number is outside this range, your data support the alternative.

If you must do a statistical analysis, be sure to do the *correct* statistical analysis.[1] Consult a statistics book or a statistician to determine which test you must use in your experiment. When comparing two means, you'll probably need only to use Student's *t* test, involving a simple calculation. (This test is not synonymous with what one author described as "Student's Tea Test"—the only statistical test that I know of that will quench your thirst.)

Remember that oddball data are rare but possible.

Always report the number of observations (n), the arithmetic mean ($\overline{x}$), and the standard deviation or standard error to give readers a sense of the data's variability (see below). Such information may be reported in the text or a table—for example, as 496 ± 60 (n = 9), in which the first two numbers show the mean plus or minus the standard deviation or standard error (indicate which), and the third number (in parentheses) represents the number of data points.

The standard deviation summarizes the spread of data about the mean, while the standard error estimates the accuracy of the mean, not the spread of the data. A small standard deviation or standard error indicates a uniform population.

The standard deviation tells you if an individual item is close to the mean. About 67% of a normal population is within one standard deviation, 95% within two standard deviations, and 99% within 2.6 standard deviations.

State the test of significance that you used with your data (a significance test summarizes the comparisons of your data with data for other populations). Also state the probability (P) values obtained with the test. These values indicate the confidence of your conclusion. Do this with statements such as $P < 0.001$, which means that the probability of the results being due to chance is less than one in a thousand. Traditionally, a result is not significant if the likelihood of obtaining the result by chance alone exceeds 5% ($P > 0.05$).

Use the word *significant* only to denote statistical significance. *Significant* is not synonymous with *important* or *meaningful*.

If you do no statistical analyses, be especially cautious when making conclusions about your data. Do not say that any differences between groups of measurements are significant or insignificant.

---

[1]In medical journals, about half the articles that use statistics use them incorrectly. See Glantz, S.A. 1980. "Biostatistics: How to Detect, Correct and Prevent Error in the Medical Literature." *Circulation* 61: 1–7; Gore, S.M., I.G. Jones, and E.C. Rytter. 1977. "Misuse of Statistical Methods: Critical Assessment of Articles in BMJ from January to March 1976. "*British Medical Journal* 1: 85–87.

Writing About Numbers

Follow these conventions when you include numbers in text:

Spell out one-digit numbers. Write all other numbers as the numeral except at the beginning of a sentence and as a measurement. Spell out numbers at the beginning of a sentence.

Fourteen grams of . . .

. . . was 14 times larger than . . .

Use numerals with standard units of measurement.

I added 104 g of NaCl to the water.

Express large numbers in figures or in mixed figure–word form.

$9,000,000 or $9 million

5,700,000 or 5.7 million

In the United States and Britain, scientists use commas to separate long numbers into groups of three digits. However, Europeans use periods or spaces to group these digits.

U.S. AND BRITISH USAGE: 1,354,000,000

EUROPEAN USAGE: 1 354 000 000

If two or more numbers appear in the same sentence and if one of the numbers exceeds 10, use numerals for all of the numbers.

I added 7 g of NaCl and 19 g of $CaCl_2$ to the media.

Express related numbers or amounts within a sentence entirely in figures or entirely in words unless doing so would confuse the reader.

Avoid placing next to each other two numbers referring to different things.

NO: In 1981 15 states ratified the law.

YES: Fifteen states ratified the law in 1981.

Hyphenate spelled-out cardinal numbers (twenty-one to ninety-nine) and ordinal numbers (twenty-first).

Don't present two numbers together.

two 100-watt lamps   *not*   2 100-watt lamps

Use a zero before a decimal point when the number is less than one.

0.21   *not*   .21

Use numerals to designate negative numbers, numbers with decimals, percentages, amounts of money, and parts of a book.

−40°C, −11, page 102, Figure 3, 4%, $8.60, Chapter 4, Table 8

Use the correct number of significant figures. Do not express the result of a calculation in more decimal places than in the least accurate part of the calculation. The accuracy of a calculation can't compensate for lack of accuracy in collection or recording data. A measurement reported as "5" is not synonymous with one reported as "5.000."

Don't write "approximately 123.5 ml." Just write "123.5 ml."

Remember that "not statistically significant" is not synonymous with "insignificant." Improbable does not equal never.

Centering and numbering equations make a document neater and more readable. Include short equations in the text. For example:

On Ludwig Boltzmann's tombstone is carved his formula for entropy, $S = K \log W$. Boltzmann committed suicide in 1906 at the age of 62.

Tables and Figures

Everything should be as simple as it can be, yet no simpler.
   —Albert Einstein

Good visuals can make or break a biology paper because people understand and remember images and pictures more readily than they do abstract words. Visuals can clarify images too complex to be described with words.

Biologists use two types of visuals in their writing: tables and figures. Tables are arrangements of numbers or words in rows and columns. Figures are everything else, including photographs, drawings, diagrams, and graphs.

Several principles hold true for tables and figures:

Do not include a table or figure unless you need it and will discuss it.

Do not present all of the data that you gathered. Rather, include only *representative* data.

Place the table or figure near where it is referred to in the text. Use words to draw attention to the table or figure.

Refer in the text to every figure, and use a consistent style for each figure. If possible, put only one table or figure on a page.

Label all axes and columns. If the figure is part of a paper that will be published, use large letters and numbers so that the information will be readable when the figure is reduced.

Include an informative title that explains the figure without referring to the text.

Be precise, clear, and concise. This doesn't mean that you should make the figures small; rather, ensure that each figure quickly informs readers of your message.

Present a realistic depiction of a process, organism, or object. Base this realism on detail included in the illustration.

Strive for clarity and simplicity. If your figures contain too much information, the reader will probably learn nothing. If your figures are too complex, either delete the figure or split it into two or more parts.

Plan the figures before you write the text so that they are an integral part of the article rather than an appendage tacked on as an afterthought. Figures must mesh with the text, like gears that drive a machine. If they don't, you're likely to end up with text and photos that resemble this advertisement that appeared in a Baltimore newspaper:

Figure 8–1
Why you need to coordinate your illustrations with your text.

Tables

Tables are an excellent way to summarize and emphasize material, reveal comparisons, show significance, and present absolute values where precision is important. However, tables must do more than merely "show the data." Well-designed tables summarize and point out the meaning and significance of data without interrupting the text. Use a table only if you must present repetitive data. Purposeless tables mystify readers, as does purposeless writing.

Here's how to design an effective table:

Include a clear, intelligible title and legend for each table. Number the tables consecutively.

Make the tables understandable and independent of the text. Don't force readers to refer to the text to understand a table. However, be sure that data in the table agree with those in the text.

Align the decimal points of numbers in the table.

Ensure that like elements in the table read downward. Include units in column headings and avoid using exponents in headings, for they often confuse readers: readers don't know if the data are to be, or have been, multiplied by these numbers. For example, "cpm $\times 10^3$" refers to thousands in the *Journal of Bacteriology*, whereas "cpm $\times 10^{-3}$" refers to thousands in the *Journal of Biological Chemistry*.

Check each table for internal accuracy. For example, do all of the percentages add up to 100? Place columns to be compared next to each other.

Explain data with footnotes at the end of the table.

Indicate the position of each table by adding a note in the margin of the text. Be sure that all tables are cited and in the correct order.

Do not use a table to describe only one or two tests. Describe such results in the text of the paper.

Include only significant, representative results in a table. If a variable has no effect, delete it from the table. Use tables to present important variables and data, not standard conditions.

Provide camera-ready copies of complicated tables. This will help you avoid laborious proofreading and help ensure accuracy.

Table 8–1 is an example of an effective table:

Table 8–1. Use of contraceptives in the United States*

Method	Estimated % use	% Accidental pregnancy in one year of use†
Male sterilization	15	0.15
Female sterilization	19	0.4
Oral contraceptive pill	32	3
Condom	17	12
Diaphragm + spermicide	5	18
"Rhythm" (periodic abstinence)	4	20
IUD	3	6
Contraceptive sponge	3	18
Vaginal foams, jellies	2	21

*Data from *Developing New Contraceptives: Obstacles and Opportunities.* Washington, DC: National Academy Press, 1990.
†About 89% of women using no contraceptives become pregnant within one year.

Figures

Although tables and figures have the same goals as language—namely, to communicate quickly, precisely, and clearly—the kind of figure that you use must be appropriate to your concept and purpose:

Figure	Best Illustrates
bar graph ("histogram")	comparisons and relative quantities; emphasizes high–low comparisons; excellent for displaying tabular data when no mathematical relationship exists between the variables; meaningless for small numbers or small differences.

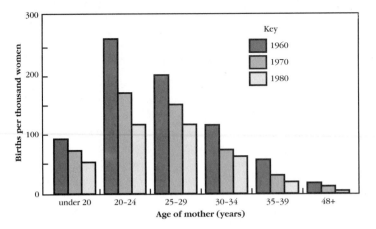

Figure 8–2

America's falling birthrate. This graph shows the number of births per thousand women in various age groups during 1960, 1970, and 1980. The number of births in all age groups decreased from 1960 to 1980, while the average age of the mother increased. Both of these trends slow population growth.

Figure	Best Illustrates
line graph	trends and relationships between variables to be compared over time

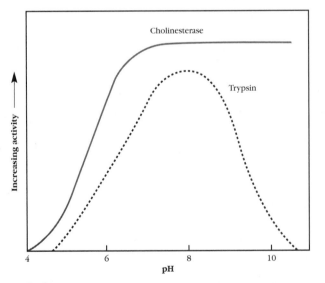

Figure 8–3

The effect of pH on the activity of trypsin and cholinesterase in mice. Trypsin is a digestive enzyme that hydrolyzes proteins in the small intestine. Trypsin is most active at pH 8, the pH in the intestine. Cholinesterase, which hydrolyzes substances important for the function of the nervous system, functions best at any basic pH. However, cholinesterase loses activity when the pH becomes acidic.

Figure	Best Illustrates
circle graph ("pie chart")	portions of a whole; makes abstract percentages appear visually as parts of the whole; poor for showing differences between small percentages; if you use a pie chart, include no more than six slices

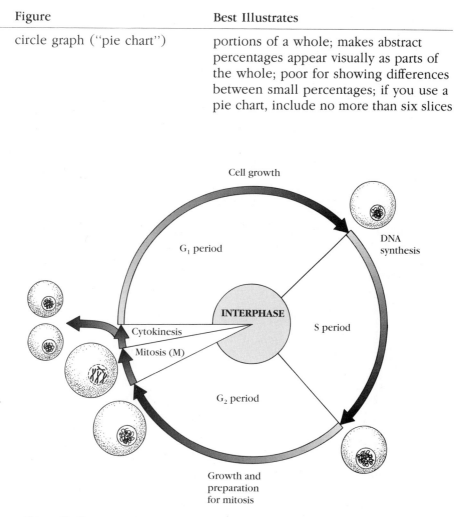

Figure 8–4

The cell cycle of a typical cell. The two gap-periods (G_1 and G_2) are separated by the synthesis (S) period, when the chromosomes replicate. Cellular division (cytokinesis) follows mitosis (nuclear division).

Figure	Best Illustrates
map	locations

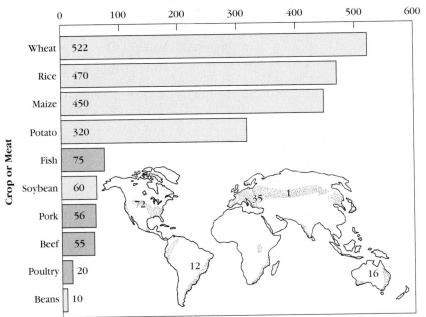

World food production in 1987 (millions of tons)

Figure 8–5

Annual world production of major foods. The four most important foods are shown at the top of the graph. The numbers on the map show 1987 grain exports in millions of tons. Several grain-growing areas lack numbers, meaning either that they produce just enough grain for their own use (India and China) or that they import grain (Africa, Latin America, and Eastern Europe). North America produces most of the grain shipped to food-importing nations.

Figure	Best Illustrates
line drawing	simple part of a process or an enlargement to show detail

Figure 8–6
The effect of skin temperature on fur color in the Himalayan rabbit. Black fur grows on the body having skin temperatures less than 33°C. If the fur is shaved from a warmer part of the body and an ice pack is applied while the fur grows back, the new fur is also black.

Figure	Best Illustrates
photograph	how something looks

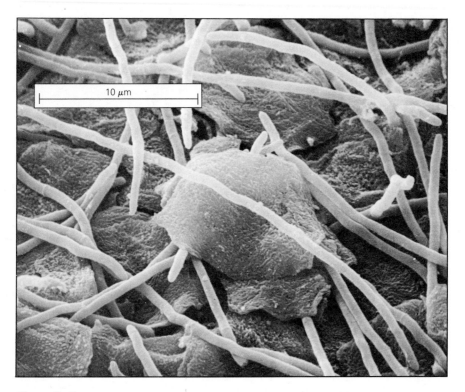

Figure 8–7

Fungal infection of human skin. Slender, hair-like hyphae grow over and under flat epidermal cells of the skin. ×4,400. Biphoto Associates

Graph

A graph is a pictorial table that allows fast comparisons of relationships or trends. As such, an effective graph illustrates trends, patterns, or relationships, promotes understanding, and suggests interpretations of data. Graphs do not add any importance or truth to data; they merely present data in pictorial form. Use a graph only if it is difficult to discuss your data in the text. Include no more data than needed to make your point.

The horizontal axis (abscissa) of a graph is called the *x*-axis, and the vertical axis (ordinate) is called the *y*-axis. The *x*-axis describes the *independent vari-*

able, which is the variable that's selected or controlled. Examples of independent variables include time, temperature, and the protein content of a diet. The *y*-axis describes the *dependent variable*, which is what was measured in response to the independent variable. Examples of dependent variables include growth rate, weight, and curvature.

Here's how to design an effective graph:

Label both axes and point the scribes (the unit marks on the axis) inward. Be sure that numbers on the *y*-axis are upright, and indicate breaks in the axis with separated lines.

Title the graph. An example of an effective title of a graph would be "The influence of temperature on root elongation."

Use the entire area of the graph. Do not bunch data at one end of the graph.

Use appropriate, evenly spaced markings on axes. Avoid empty spaces on the graph by not extending the axes unnecessarily.

If the independent variable is continuous (for example, time or temperature), use a line graph. If the independent variable is not continuous, such as when you're comparing the heights of seedlings of lettuce and strawberries, use a bar graph.

Use common symbols such as ■, □, ●, and ○ for data points. Include a key for each symbol in the graph. Use symbols consistently: Use the same symbols in each graph.

If you plot more than one set of data on a graph, use different lines for each set of data. Label each line clearly. Plot no more than five lines per graph.

Do not extrapolate beyond the data points.

Box the graph. This makes it easier for readers to estimate values on the right side of the graph.

Unless you're writing a term paper or lab report, print the legend for the graph on a separate page.

Indicate the standard-error bars for each mean, with vertical lines extending up and down from each point. These "error bars" tell readers how reliably each mean estimates the population mean. You can also communicate this information by plotting 95% confidence intervals at each mean. To show variability within each sample, plot the standard deviation or standard error. Specify in the legend what statistic you've plotted.

How you should connect the data points on the graph depends on your data and how they were obtained, analyzed, and interpreted.

To show a trend, use a smooth line. Most of the data points will touch or be close to the line. Sampling error accounts for why all points do not lie on the line. This is shown in Fig. 8–8:

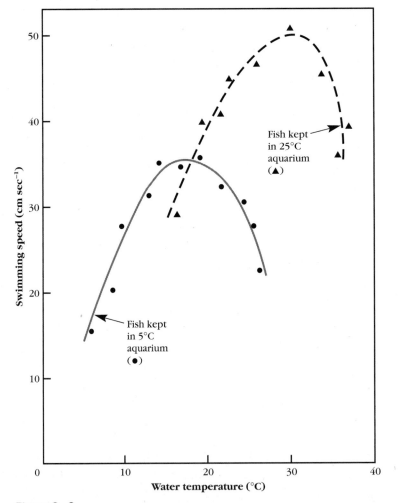

Figure 8–8

How water temperature affects swimming speed in a goldfish. There is an optimal water temperature for swimming and a range of temperatures the goldfish can tolerate. By gradually changing the water temperature, the goldfish can tolerate higher or lower temperatures than it could had it been moved to the new environment abruptly.

To show a mathematical relationship between the two variables, include a statistically derived "best fit" line. In the legend, define the regression coefficient for the line. This is shown in Fig. 8–9:

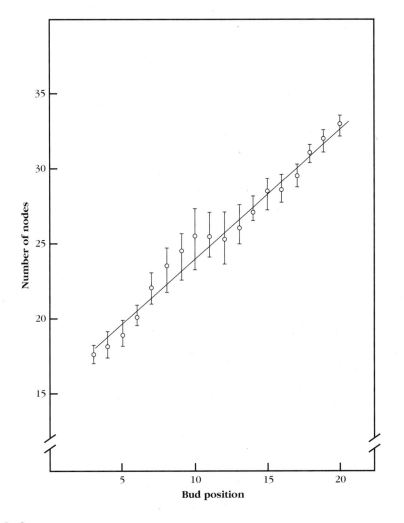

Figure 8–9

Position-dependent growth of axillary buds of *Nicotiana tabacum*. I've plotted the average number of nodes ± standard deviation produced by four buds at each position. Buds were numbered basipetally with bud number 1 being the first axillary bud immediately below the inflorescence. The regression coefficient (R^2) for the line equals 0.92.

To show high–low patterns, connect the data points with straight lines. The gaps between these points will not be due to sampling error but rather to meaningful fluctuations. For example, Figure 8–10 shows how the size of a population of fruit flies changes when kept in a closed container and fed a constant diet of protein. The changes in the population are biologically important. This is shown in Fig. 8–10:

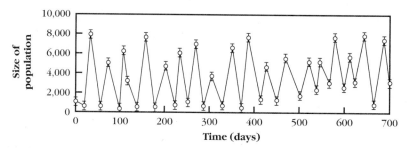

Figure 8–10

Oscillation in the size (± standard deviation) of a population of *Drosophila melanogaster*. Flies were kept in a closed container and fed a constant diet of protein.

Many computer programs such as *Excel, Chart,* and *SigmaPlot* can generate a variety of graphs from data. Most of these graphs, when printed on a laser printer, are acceptable by journals for publication.

Line Drawings

A line drawing is an artist's concept of an object or concept constructed with lines, as shown in Figure 8–11.

Line drawings are usually drawn with black India ink and include no shades of gray. They should not show everything; if you want to show everything, use a photograph. Use drawings to omit extraneous details such as with a cutaway depicting the inner workings of a device. Make the drawing as simple as possible so that it won't hide the key ideas.

Make drawings to scale and delete excess information. Make the drawings twice their final size to ensure accuracy and detail. Lines should be at least 0.3 mm thick.

Photographs

When it comes to showing precisely how an object appears, nothing is better than the realism of a photograph. Photographs show exact details but can't speak for themselves. Similarly, photographs can't lie, but they can mislead. Therefore,

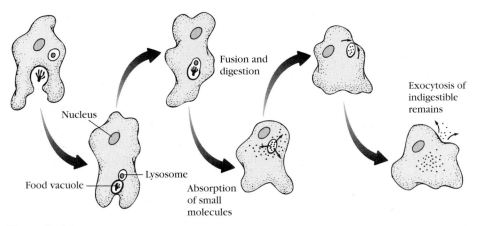

Nucleus

Food vacuole

Lysosome

Fusion and digestion

Absorption of small molecules

Exocytosis of indigestible remains

Figure 8–11

Phagocytosis, a type of endocytosis. An amoeba engulfs its prey, and part of the plasma membrane pinches off to form a food vacuole in the cell. This food vacuole fuses with a lysosome, a membrane-bound organelle full of digestive enzymes. The small food molecules formed by digestion are absorbed into the cytoplasm before the amoeba expels the indigestible remains of the prey by exocytosis.

be sure that each photograph represents the information you are trying to communicate. Like illustrations, photographs are not proof; the burden of proof lies with the author. However, photographs are more credible than drawings because photographs show that the object actually exists and was not constructed from an artist's imagination.

Contrary to what many scientists think, not all pictures are worth 1,000 words. However, you can help make them so by keeping these things in mind:

Take photographs so that they highlight the purpose of the photograph. Be sure that each photograph is in excellent focus, has good contrast, and has a background that doesn't detract from the subject. Mount the photographs into a montage (also called a photographic "plate") according to the journal's instructions to authors.

When taking photographs for a paper, fill the frame of the photograph with the subject. Also, remember what many consider Ansel Adams's best piece of advice: "If you want to take a good picture, take a lot of pictures."

Photographs should be glossy black-and-white prints. These photographs are referred to as "halftones" because, unlike line drawings, they include shades of gray. Submit color photographs only if the journal accepts color photographs and you are willing to pay for printing the color photographs. Ask the journal's managing editor about these costs.

Attach an adhesive sticker to the back of each photograph to indicate the top and the figure number of the photograph. Similarly, indicate the approximate location of each photograph with a note in the margin of the text.

Use rub-on letters and arrows to highlight important features of the photograph and letters to label other important parts. Include a scale-marker on each photograph.

Include a legend describing the important aspects of each photograph. If submitting the photograph for publication, type the legend on a separate page of paper. The figure and the legend are dealt with differently by publishers: Legends go to typesetting, while photographs are handled by photoengravers. Figure 8–12 is an example of a well-composed photograph.

Figure 8–12
Different eyes see different things. Human eyes see a marsh marigold with no markings (*left*). Insects see large patches on the same flower (*right*, taken with ultraviolet-sensitive film) because their eyes react to ultraviolet light, whereas our eyes do not. Biphoto Associates

A Final Check

Just as effective illustrations and photographs can dramatically improve the clarity and impact of your paper, so too can poor illustrations and photographs damage your credibility and discredit your work. Therefore, before turning in a paper or submitting a paper for publication, check the following:

Check every illustration against your original artwork. Did the artist introduce any mistakes?

Do the figure numbers correspond with those cited in the text?

Are the words, letters, and abbreviations on the figures consistent with those used in the text?

Is the figure drawn to scale? Is this scale marked on the illustration?

Does the figure contain too much information?

Are the units of measurement clearly marked on all axes and scales?

Should any photograph be replaced by a line drawing?

Should any line drawing be replaced by a photograph?

Is a concise, understandable legend included with each figure?

Do the photographs show what I want them to show? Are they in good focus?

Do all the illustrations mesh with the text?

Do not sacrifice truth for appeal. However, make sure that each photograph or illustration is aesthetically appealing and professionally done. Editors are puzzled by authors who, after spending $100,000 and several months to gather data, then refuse to spend $10 to have a draftsperson prepare a professional-quality graph.

J.B.S. Haldane

On Being the Right Size

J.B.S. Haldane (1892–1964) is famous for mathematically analyzing heredity and estimating the rate of genetic mutation. In recognition of his work, in 1932 Haldane received the Darwin Medal of the Royal Society for starting "the modern phase of study of the evolution of living populations." Haldane led a colorful life. In his studies of human physiology, especially those of sunstroke and "the bends," he was often his own experimental animal. In the 1930s, he became a Communist and helped aid Nazi refugees. Haldane spent his last years in self-exile in India out of dissatisfaction with British policies.

Most of Haldane's essays were first published in periodicals. Like many great scientists, Haldane wrote such essays because he believed that the "public has a right to know what is going on inside the labs, for some of which they pay." Today, scientists often cite Haldane's "On Being the Right Size" as an example of how science can change how we look at things. This and many other essays of Haldane's are published in *Possible Worlds* (1928) and were written "in the intervals of research work and teaching, to a large extent in railway trains."

The most obvious differences between different animals are differences of size, but for some reason the zoologists have paid singularly little attention to them. In a large textbook of zoology before me I find no indication that the eagle is larger than the sparrow, or the hippopotamus bigger than the hare, though some grudging admissions are made in the case of the mouse and the whale. But yet it is easy to show that a hare could not be as large as a hippopotamus, or a whale as small as a herring. For every type of animal there is a most convenient size, and a large change in size inevitably carries with it a change of form.

Let us take the most obvious of possible cases, and consider a giant man sixty feet high—about the height of Giant Pope and Giant Pagan in the illustrated *Pilgrim's Progress* of my childhood. These monsters were not only ten times as high as Christian, but ten times as wide and ten times as thick, so that their total weight was a thousand times his, or about eighty to ninety tons. Unfortunately the cross sections of their bones were only a hundred times those of Christian, so that every square inch of giant bone had to support ten times the weight borne by a square inch of human bone. As the human thigh-bone breaks under about ten times the human weight, Pope and Pagan would have broken their thighs every time they took a step. This was doubtless why they were sitting down in the picture I remember. But it lessens one's respect for Christian and Jack the Giant Killer.

To turn to zoology, suppose that a gazelle, a graceful little creature with long thin legs, is to become large, it will break its bones unless it does one of two things. It may make its legs short and thick, like the rhinoceros, so that every pound of weight has still about the same area of bone to support it. Or it can compress its body and stretch out its legs obliquely to gain stability, like the giraffe. I mention these two beasts because they happen to belong to the same order as the gazelle, and both are quite successful mechanically, being remarkably fast runners.

Gravity, a mere nuisance to Christian, was a terror to Pope, Pagan, and Despair. To the mouse and any smaller animal it presents practically no dangers. You can drop a mouse down a thousand-yard mine shaft; and, on arriving at the bottom, it gets a slight shock and walks away, provided that the ground is fairly soft. A rat is killed, a man is broken, a horse splashes. For the resistance presented to movement by the air is proportional to the surface of the moving object. Divide an animal's length, breadth, and height each by ten; its weight is reduced to a thousandth, but its surface only to a hundredth. So the resistance to falling in the case of the small animal is relatively ten times greater than the driving force.

An insect, therefore, is not afraid of gravity; it can fall without danger, and can cling to the ceiling with remarkably little trouble. It can go in for elegant and fantastic forms of support like that of the daddy-longlegs. But there is a force which is as formidable to an insect as gravitation to a mammal. This is surface tension. A man coming out of a bath carries with him a film of water of about one-fiftieth of an inch in thickness. This weighs roughly a pound. A wet mouse has to carry about its own weight of water. A wet fly has to lift many times its own weight and, as everyone knows, a fly once wetted by water or any other liquid is in a very serious position indeed. An insect going for a drink is in as great danger as a man leaning out over a precipice in search of food. If it once falls into the grip of the surface tension of the water—that is to say, gets wet—it is likely to remain so until it drowns. A few insects, such as water-beetles, contrive to be unwettable; the majority keep well away from their drink by means of a long proboscis.

Of course tall land animals have other difficulties. They have to pump their blood to greater heights than a man, and, therefore, require a larger blood pressure and tougher blood-vessels. A great many men die from burst arteries, especially in the brain, and this danger is presumably still greater for an elephant or a giraffe. But animals of all kinds find difficulties in size for the following reason. A typical small animal, say a microscopic worm or rotifer, has a smooth skin through which all the oxygen it requires can soak in, a straight gut with sufficient surface to absorb its food, and a single kidney. Increase its dimensions tenfold in every direction, and its weight is increased a thousand times, so that if it is to use its muscles as efficiently as its miniature counterpart, it will need a thousand times as much food and oxygen per day and will excrete a thousand times as much of waste products.

Now if its shape is unaltered its surface will be increased only a hundredfold, and ten times as much oxygen must enter per minute through each square millimetre of skin, ten times as much food through each square millimetre of intestine. When a limit is reached to their absorptive powers their surface has to be increased by some special device. For example, a part of the skin may be drawn out into tufts to make gills or pushed in to make lungs, thus increasing the oxygen-absorbing surface in proportion to the animal's bulk. A man, for example, has a hundred square yards of lung. Similarly, the gut, instead of being smooth and straight, becomes coiled and develops a velvety surface, and other organs increase in complication. The higher animals are not larger than the lower because they are more complicated. They are more complicated because they are larger. Just the same is true of plants. The simplest plants, such as the green algae growing in stagnant water or on the bark of trees, are mere round cells. The higher plants increase their surface by putting out leaves and roots. Comparative anatomy is largely the story of the struggle to increase surface in proportion to volume.

Some of the methods of increasing the surface are useful up to a point, but not capable of a very wide adaptation. For example, while vertebrates carry the oxygen from the gills or lungs all over the body in the blood, insects take air directly to every part of their body by tiny blind tubes called tracheae which open to the surface at many different points. Now, although by their breathing movements they can renew the air in the outer part of the tracheal system, the oxygen has to penetrate the finer branches by means of diffusion. Gases can diffuse easily through very small distances, not many times larger than the average length travelled by a gas molecule between collisions with other molecules. But when such vast journeys—from the point of view of a molecule—as a quarter of an inch have to be made, the process becomes slow. So the portions of an insect's body more than a quarter of an inch from the air would always be short of oxygen. In consequence hardly any insects are much more than half an inch thick. Land crabs are built on the same general plan as insects, but are much clumsier. Yet like ourselves they carry oxygen around in their blood, and are therefore able to grow far larger than any insects. If the insects had hit on a plan for driving air through their tissues instead of letting it soak in, they might well have become as large as lobsters, though other considerations would have prevented them from becoming as large as man.

Exactly the same difficulties attach to flying. It is an elementary principle of aeronautics that the minimum speed needed to keep an aeroplane of a given shape in the air varies as the square root of its length. If its linear dimensions are increased four times, it must fly twice as fast. Now the power needed for the minimum speed increases more rapidly than the weight of the machine. So the larger aeroplane, which weighs sixty-four times as much as the smaller, needs one hundred and twenty-eight times its horsepower to keep up. Applying the same principle to the birds, we find that the limit to their size is soon reached. An angel whose muscles developed no more power weight for weight than those of an eagle or a pigeon would require a breast projecting for about four feet to house the muscles engaged in working its wings, while to economize in weight, its legs would have to be reduced to mere stilts. Actually a large bird such as an eagle or kite does not keep in the air mainly by moving its wings. It is generally to be seen soaring, that is to say balanced on a rising column of air. And even soaring becomes more and more difficult with increasing size. Were this not the case eagles might be as large as tigers and as formidable to man as hostile aeroplanes.

But it is time that we pass to some of the advantages of size. One of the most obvious is that it enables one to keep warm. All warm-blooded animals at rest lose the same amount of heat from a unit area of skin, for which purpose they need a food-supply proportional to their surface and not to their weight. Five thousand mice weigh as much as a man. Their combined surface and food or oxygen consumption are about seventeen times a man's. In fact a mouse eats about one quarter its own weight of food every day, which is mainly used in keeping it warm. For the same reason small animals cannot live in cold countries. In the arctic regions there are no reptiles or amphibians, and no small mammals. The smallest mammal in Spitzbergen is the fox. The small birds fly away in winter, while the insects die, though their eggs can survive six months or more of frost. The most successful mammals are bears, seals, and walruses.

Similarly, the eye is a rather inefficient organ until it reaches a large size. The back of the human eye on which an image of the outside world is thrown, and which corresponds to the film of a camera, is composed of a mosaic of "rods and cones" whose diameter is little more than a length of an average light wave. Each eye has about a half a million, and for two objects to be distinguishable their images must fall on separate rods or cones. It is obvious that with fewer but larger rods and cones we should see less distinctly. If they were twice as broad two points would have to be twice as far apart before we could distinguish them at a given distance. But if their size were diminished and their number increased we should see no better. For it is impossible to form a definite image smaller than a wave-length of light. Hence a mouse's eye is not a small-scale model of a human eye. Its rods and cones are not much smaller than ours, and therefore there are far fewer of them. A mouse could not distinguish one human face from another six feet away. In order that they should be of any use at all the eyes of small animals have to be much larger in proportion to their bodies than our own. Large animals on the other hand only require relatively small eyes, and those of the whale and elephant are little larger than our own.

For rather more recondite reasons the same general principle holds true of the brain. If we compare the brain-weights of a set of very similar animals such as the cat, cheetah, leopard, and tiger, we find that as we quadruple the body-weight the brain-weight is only doubled. The larger animal with proportionately larger bones can economize on brain, eyes, and certain other organs.

Such are a very few of the considerations which show that for every type of animal there is an optimum size. Yet although Galileo demonstrated the contrary more than three hundred years ago, people still believe that if a flea were as large as a man it could jump a thousand feet into the air. As a matter of fact the height to which an animal can jump is more nearly independent of its size than proportional to it. A flea can jump about two feet, a man about five. To jump a given height, if we neglect the resistance of air, requires an expenditure of energy proportional to the jumper's weight. But if the jumping muscles form a constant fraction of the animal's body, the energy developed per ounce of muscle is independent of the size, provided it can be developed quickly enough in the small animal. As a matter of fact an insect's muscles, although they can contract more quickly than our own, appear to be less efficient; as otherwise a flea or grasshopper could rise six feet into the air.

On what does the effect of scale depend?

Haldane relies heavily on examples to make his point. What kinds of examples does he use and how does his ordering of them sharpen his argument?

All objects and organisms have their proper scale and absolute magnitudes. What other concepts besides scale might be the subjects of an essay like Haldane's?

Exercises

1. Examine the data shown in each of the figures and tables in this chapter. Write an essay about each set of data. Do not repeat the data but discuss what they mean.

2. Examine the data shown in Exercise 8 of Chapter 2. What is the most effective way of displaying these data? Why? Use the space below to plot the data.

3. Discuss the ideas included in these quotations:

> Population, when unchecked, increases in geometrical ratio. Subsistence increases only in an arithmetical ratio. —Thomas Robert Malthus

> The most intensively social animals can only adapt to group behavior. Bees and ants have no option when isolated, except to die. There is really no such creature as a single individual; he has no more life of his own than a cast-off cell marooned from the surface of your skin. —Lewis Thomas

CHAPTER NINE

Oral Presentations

A speech is a solemn responsibility. The man who makes a bad thirty-minute speech to two hundred people wastes only a half hour of his own time. But he wastes one hundred hours of the audience's time—more than four days—which should be a hanging offense.
> —Jenkin Lloyd Jones

Listen up, because I've got nothing to say and I'm only going to say it once.
> —Yogi Berra

Make sure you have finished speaking before your audience has finished listening.
> —Dorothy Sarnoff

Be sincere; be brief; be seated.
> —Franklin D. Roosevelt

You're probably puzzled by a chapter entitled "Oral Presentations" in a book entitled *Writing to Learn Biology*. I've included this chapter because talking effectively about biology, like writing effectively about biology, is important to a biologist's career. Most scientific meetings include "paper sessions" in which scientists talk about their work.

The objective of a talk at a scientific meeting is to communicate your discovery to your audience. To do this, you must understand the nature of your audience and capture their interest with logic, effective graphics, well-organized ideas, and simplicity. Since your audience listens to and looks at you, you must concentrate on what you say and do in your talk.

Here's how to present an effective talk:

Know your audience. Gear your talk to their needs and interests.

Unlike writing, oral presentations are designed for immediate consumption. Therefore, present fewer data and stress generalizations. Define all terms in the introduction of your talk and don't hesitate to restate your conclusions.

Mix your discussions of results and discussion. Since listeners, unlike readers, have no way to re-examine what came before, having distinct "Results" and "Discussion" sections is usually not effective.

Capture the group's attention by politely saying "Good morning" or "Good afternoon" at the beginning of your talk. Don't start your talk by saying that you can't compress a year's worth of work into the 15 minutes you've been allocated. You've not been asked to compress any information; you're expected only to communicate well.

Project restrained self-confidence by projecting your voice, punctuating your words with animation, pronouncing each word crisply, and avoiding fillers such as "okay," "you know," and "uhhh."

Practice your talk several times to ease your nerves about giving it before a group. Face your audience and don't be a sleeping pill; show your enthusiasm for your work.

State the objectives of your work clearly, concentrate on concepts, and eliminate details. People can learn the details of your work by reading your papers or asking you questions after your talk.

Focus on your results and conclusions. Discuss the purpose, rationale, and conclusions of your work as you go. Emphasize these points by repeating them in different ways. Use evidence and logic to lead readers to your conclusion.

Use large-print note cards, but don't read your talk. Rather, maintain eye contact with your audience to convey sincerity and conviction. To help avert a case of stage fright, memorize the first few sentences of your talk.

Obey the time limit. Most presenters are given 15 minutes for a talk. Use about 2 minutes to introduce your talk, 11 minutes for the body of your talk, and the remaining 3 minutes for conclusions and questions. There are few things more disrespectful to your peers than exceeding the time allocated for your presentation. Practice your talk so that you do not exceed your allotted time.

Maintain good posture. Avoid distracting mannerisms such as rocking back and forth, picking your nose, or grabbing your crotch as you talk.

Speak slowly and clearly at about 100 words per minute. At that rate, you'll cover a double-spaced page of text in about 2.5 minutes. Use short, forceful sentences. If a microphone is available, stay about 15 cm from the microphone. If a microphone isn't available, speak loudly and project your voice as if you're talking to someone sitting in the back of the room.

Don't turn your back on the audience. This is impolite and reduces the projection of your voice.

At the end of your talk be prepared to answer questions about your work. When asked questions, first repeat the question to the group. Then answer the question directly. If you don't know the answer, just say "I don't know."

Announce the end of your talk by saying something like "Thank you" or "I'll be glad to answer questions." Don't just abruptly stop talking; people won't know if they should applaud or if your mind's just gone blank.

Photographic slides, handouts, and overhead transparencies are critical for most presentations, especially those given at meetings. Be sure that they're done professionally and that data aren't crowded onto any of the slides. When practicing your talk, sit at the back of a large hall while your slides are projected so that you can check them for accuracy and readability.

Prepare slides, transparencies, and handouts expressly for your talk. If you use those made from tables and figures of a manuscript, you'll probably end up telling the audience things such as "Ignore everything on this slide except this column." Design your talk so that the audience does not have to ignore anything you say or show.

Allow about 1 minute per slide or overhead transparency. Give the audience time to look at each slide and then help the audience interpret each slide. You and your audience should be able to understand a slide in less than 5 seconds. Title each slide and do not read each word on a slide to the audience. Rather, use slides to supplement what you say. Cramming too many slides into your talk will only frustrate your audience.

Do not show a slide if it includes too much detail or anything that is irrelevant to your talk. Use one slide to convey one idea, not detailed data. Make that idea brief, clear, and simple so that it can be understood quickly. Be sure that all slides have fewer than eight lines and four columns.

Do not show a table if it has too many numbers or if the numbers are so small that people cannot read them. If the slide is not readable when held at arm's length, it will not be readable when projected from the back of a conference hall.

Remove each slide after it has served its purpose so that it is not shown while you are trying to interest your audience in something else.

Make sure that all slides are in their proper orientation so that they will be projected right-side-up.

Use labels to focus the attention of the audience on the important parts of the slide.

Find a pointer before you speak. Don't wait for the lights to go off before you start fumbling for a pointer.

Use slide carousels that hold 80 rather than 140 slides (the 140-capacity carousels are more likely to jam). However, if the projector jams during your talk, keep talking. If you stop talking, you quickly lose the attention of the audience.

Plan slides so that their longest dimension is horizontal when projected. It's hard to see vertically oriented slides in rooms having low ceilings.

Use uppercase and lowercase letters, especially if your slide has more than about five words.

Start and end your talk with an opaque slide. Don't blind your audience with an empty frame.

Carry your slides, transparencies, and handouts with you if you travel to give your talk. Don't pack your slides in your checked baggage unless you want to risk having the airline lose or destroy them.

Exercises

1. Prepare a 10-minute talk about any of the subjects that you've written about in this course. Be prepared to give the talk to your class.

2. Prepare oral and written essays about the same subject. How are they different? How are they similar?

E P I L O G

Using What You've Learned

We are all apprentices in a craft where no one ever
becomes a master.
    —Ernest Hemingway

What can you learn in college that will help in
being an employee? . . . There you learn the one
thing that is perhaps the most valuable for the
future employee to know. This one basic skill is the
ability to organize and express ideas in writing and
speaking.
    —Peter Drucker, *Fortune*

Good writing skills are an invaluable, but often
understated, necessity for career and business. The
person who develops these skills will certainly be
recognized early. . . .
    —Terry C. Carder, Chairman of the Board,
      Reynolds & Reynolds Information Systems

Practice, practice. Put your hope in that.
    —M.S. Merwin

If you've studied the principles discussed in this book, you're now a better writer than before: You know how to use writing to discover your ideas, how to organize those ideas into a first draft, and how to shape these ideas into a paper or laboratory report that communicates your ideas effectively. In short, you have a better understanding of the writing process and therefore can learn biology by writing about biology. This ability to write well and to use writing as a learning tool will benefit you throughout your career. Biologists and others who insist on writing poorly will envy your ability to communicate well; they'll also continue to cloak their ideas in secrecy by preferring jargon to clear writing, by sacrificing the beauty of the active voice to seem objective, and by remaining blind to imagination despite their claims of seeing "the big picture." Don't let these peoples' attitudes bother you. Insist on writing clearly. If you do, people will be impressed by your ideas rather than confused by your writing.

If you can now write clearly, coherently, and emphatically, you'll be tempted to stop trying to improve your writing. Most people would be satisfied to have accomplished so much. However, like other tools of biologists, writing is a craft that can be improved by practice and study. To show this, consider the following essays of William Osler. In 1889 Osler wrote like this:

> In a true and perfect form, imperturbability is indissolubly associated with wide experience and an intimate knowledge of the varied aspects of disease. With such advantages he is so equipped that no eventuality can disturb the mental equilibrium of the physician; the possibilities are always manifest, and the courts of action clear. From its very nature this precious quality is liable to be misinterpreted, and the general accusation of hardness, so often brought against the profession, has here its foundation.[1]

Although this essay has a worthwhile message, Osler uses too many big words and phrases such as *imperturbability is indissolubly associated* that hinder readers. Three years later Osler wrote like this about one of his favorite topics, the need for medical students to work with patients:

> I would fain dwell upon many other points in the relation of the hospital to the medical school — on the necessity of ample, full and prolonged clinical instruction, and on the importance of bringing the student and the patient into close contact, not through the cloudy knowledge of the wards; on the propriety of encouraging the younger men as instructors and helpers in ward work; and on the duty of hospital physicians and surgeons to contribute to the advance of their art. . . .[2]

Although this essay is simpler and uses fewer big words than did his earlier essay, it still does not say things simply. Moreover, it lacks grace, has no rhythm,

---

[1]Osler, W. 1932. "Aequanimitas." In *Aequanimitas* (3rd ed.). Philadelphia: Blakiston. p. 315.
[2]Osler, W. 1932. "Teacher and Student." In *Aequanimitas* (3rd ed.). Philadelphia: Blakiston. p. 31.

and sounds pompous. Compare this with what Osler wrote more than a decade later:

> Ask any physician of twenty years' standing how he has become proficient in his art, and he will reply, by constant contact with disease; and he will add that the medicine he learned in the schools was totally different from the medicine he learned at the bedside. The graduate of a quarter of a century ago went out with little practical knowledge, which increased only as his practice increased. In what may be called the natural method of teaching the student begins with the patient, continues with the patient, and ends with the patient, using books and lectures as tools, as means to an end. The student starts, in fact, as a practitioner. . . .[3]

What a difference! Unlike in the previous essays, here Osler relies on strong verbs and simple words (most have only one or two syllables), avoids useless repetition, and uses few adjectives. Consequently, this essay is forceful, clear, and easy to understand.

Another excellent way to improve your writing is to read good writing. Just as a child isolated by deafness has problems learning to speak, so too does a writer who does not read have problems in writing. Therefore, take time to read the works of classical and modern writers. Don't be afraid to model your writing after your favorite writers; after all, Bach and Picasso used models, and you'll also benefit by studying the works of great writers.

Reading great biology books will help you enjoy and appreciate how to use words effectively. Here are some good books by biologists to start with:

> Rachel Carson's *Silent Spring* heralded the age of ecological awareness by describing the hidden effects of human dominance of nature.

> Lewis Thomas's *The Lives of a Cell: Notes of a Biology Watcher* and *The Medusa and the Snail* are compilations of his essays from *The New England Journal of Medicine*. Thomas writes about the common aspects of science so readers can follow his ideas into the world of their experiences. *The Lives of a Cell: Notes of a Biology Watcher* won the National Book Award in 1975.

> James Watson's best-seller *The Double Helix* describes an exhilarating race to solve a scientific puzzle. His book, along with C.P. Snow's *The Search*, will show you that scientists have their share of vanity, greed, lust, sloth, and indecisiveness.

> Jane Goodall's *In The Shadow of Man* describes Goodall's classic research on primate behavior and shows the power of science to enthrall.

> Stephen Jay Gould's books and *Natural History* essays are popular because Gould realizes that everyone likes a good story. Gould views

[3]Osler, W. 1932. "The Hospital as a College." In *Aequanimitas* (3rd ed.). Philadelphia: Blakiston. p. 315.

science as a man trying to understand forces much greater than he is. His *The Mismeasure of Man* won the National Book Critics' Award in 1981, and his *The Panda's Thumb* won the American Book Award for Science in 1981.

Other great books about biology include Ernst Mayr's *The Growth of Biological Thought*, Thomas Kuhn's *The Structure of Scientific Revolutions*, Paul De Kruif's *Microbe Hunters*, and E.O. Wilson's *On Human Nature*. Although you'll like these books, don't restrict your reading to books written only by biologists or other scientists. Also read the works of writers such as E.B. White, Ernest Hemingway, Ralph Waldo Emerson, Rebecca West, William James, Mark Twain, Henry David Thoreau, Robert Louis Stevenson, Thornton Wilder, John Steinbeck, James Thurber, Barbara Tuchman, Art Buchwald, and H.L. Mencken. Reading these writers' works will get the shapes and rhythms of good writing into your head. They'll also help you more than reading about good writing and are much more entertaining. Moreover, you'll be surprised by the scientific insights of many of these writers. For example, Oliver Wendell Holmes became famous for his poem "Old Ironsides." However, Holmes was also a physician who coined the term *anesthetic* and wrote extensively about biological topics such as the contagiousness of puerperal fever. He was known as so astute a scientific reasoner that his British colleague Arthur Conan Doyle named the incomparable and eccentric detective Sherlock Holmes after him.

Finally, remember that what you write will be used by other biologists and will constitute much of your legacy as a biologist. Give it great care.

APPENDIX ONE

Useful References

Figures and Tables

Allen, A. 1977. *Steps Toward Better Scientific Illustrations.* 2nd edition. Lawrence, KS: Allen Press, Inc. An excellent discussion of illustrations and photographs.

CBE Scientific Illustration Committee. 1988. *Illustrating Science: Standards for Publication.* Bethesda, MD: Council of Biology Editors.

Cleveland, W.S. 1985. *The Elements of Graphing Data.* Monterey, CA: Wadsworth Advanced Books and Software. This book for advanced readers includes many good and bad examples from the scientific literature.

Selby, P.H. 1976. *Interpreting Graphs and Tables.* New York: John Wiley and Sons. A self-teaching manual aimed at beginners.

Tufte, E.R. 1983. *The Visual Display of Quantitative Information.* Cheshire, CT: Graphics Press. A thorough book aimed at an advanced audience.

Writing, Grammar, and Punctuation

Barnet, S., and M. Stubbs. 1986. *Practical Guide to Writing.* 5th edition. Boston: Little, Brown and Co.

Barzun, J. 1985. *Simple & Direct: A Rhetoric for Writers.* New York: Harper & Row. Includes 20 principles of clear writing.

Bates, Jefferson D. 1990. *Writing With Precision.* 2nd edition. Herndon, VA: Acropolis Books Ltd.

Brittain, Robert. 1981. *A Pocket Guide to Correct Punctuation.* Woodbury, NY: Barron's Educational Series.

Chicago Manual of Style, The. 1982. 13th edition. Chicago: University of Chicago Press.

The standard reference and source book for writers. Covers a wide range of scientific writing and is an excellent supplement to guides devoted to scientific writing.

Collinson, Diane, et al. *Plain English*. 1977. Milton Keynes, England: The Open University Press. A self-teaching text that contains many exercises, tests, and clear explanations.

Cook, Claire K. 1985. *Line by Line: How to Improve Your Own Writing*. Boston, MA: Houghton Mifflin.

Corbett, Edward P.J. 1982. *The Little Rhetoric and Handbook*. Glenview, IL: Scott, Foresman, and Co. One of the few "writing books" that relates writing to reasoning.

Delton, Lucy. 1985. *The 29 Most Common Writing Mistakes and How to Avoid Them*. Cincinnati: Writer's Digest Books.

Flower, L. 1985. *Problem-Solving Strategies for Writing*. 2nd edition. New York: Harcourt Brace Jovanovich.

Fowler, H.W. 1965. *A Dictionary of Modern English Usage*. 2nd edition. Oxford: Oxford University Press. The most reliable reference for matters of usage.

Gordon, Karen E. 1983. *The Well-Tempered Sentence*. New York: Ticknor & Fields. A short, funny survey of the rules of punctuation.

Gordon, Karen E. 1984. *The Transitive Vampire*. New York: Times Books. An entertaining, readable introduction to grammar.

Hairston, M.C. 1981. *Successful Writing*. 2nd edition. New York: W.W. Norton. Includes a thorough discussion of revising papers.

Hall, D. 1985. *Writing Well*. 5th edition. Boston: Little, Brown and Co.

Hooper, Vincent R., Cedric Gale, Ronald C. Foote, and Benjamin W. Griffith. 1982. *Essentials of English*. 3rd edition. Woodbury, NY: Barron's Educational Series.

Kaufer, D.S., C. Geisler, and C.M. Neuwirth. 1989. *Arguing from Sources: Exploring Issues through Reading and Writing*. San Diego: Harcourt Brace Jovanovich. Especially useful for learning how to develop arguments in expository writing.

Maggio, R. 1987. *The Nonsexist Word Finder: A Dictionary of Gender-Free Usage*. Phoenix: Oryx.

Martin, Phyllis. 1982. *Word Watchers Handbook: A Deletionary of the Most Abused and Misused Words*. New York: St. Martin's Press.

Roberts, Philip D. 1987. *Plain English: A User's Guide*. Middlesex, England: Penguin Books.

Strunk, William, and E.B. White. 1979. *The Elements of Style*. 3rd edition. New York: Macmillan Publishing Co. The best book ever written about writing. If you can afford only one book about writing, this is the one to buy. Read this book.

Temple, Michael. 1978. *A Pocket Guide to Correct Punctuation*. Woodbury, NY: Barron's Educational Series.

Walsh, J. Martyn, and Anna K. Walsh. 1982. *Plain English Handbook: A Complete Guide to Good English*. 8th edition. New York: Random House.

Williams, Joseph M. 1990. *Style*. Chicago: University of Chicago Press.

Woolston, Donald C., Patricia A. Robinson, and Gisela Kutzbach. 1988. *Effective Writing Strategies for Engineers and Scientists*. Chelsea, MI: Lewis Publishers.

Writing with a Computer

Hult, C., and J. Harris. 1987. *A Writer's Introduction to Word Processing*. Belmont, CA: Wadsworth, Inc. A practical, start-from-scratch guide to using word-processing in your writing.

Mitchell, Joan P. 1987. *Writing with a Computer*. Boston: Houghton Mifflin.

Scientific and Technical Writing

Alley, Michael. 1987. *The Craft of Scientific Writing*. Englewood Cliffs, NJ: Prentice-Hall.

Bates, Jefferson D. 1988. *Writing with Precision*. Washington: Acropolis Books Ltd.

Bly, Robert W., and Gary Blake. 1982. *Technical Writing: Structure, Standards, and Style*. New York: McGraw-Hill.

Day, Robert A. 1983. *How to Write and Publish a Scientific Paper*. 2nd edition. Philadelphia: ISI Press. A short, witty book containing much practical advice about preparing a paper for publication.

Gray, P. 1982. *The Dictionary of the Biological Sciences*. New York: Krieger. Gives definitions, spelling, pronunciation, and usage of technical and scientific terms.

Huth, Edward J. 1982. *How to Write and Publish Papers in the Medical Sciences*. Philadelphia: ISI Press.

Huth, Edward J. 1987. *Medical Style and Format: An International Manual for Authors, Editors, and Publishers*. Philadelphia: ISI Press.

King, L.S. 1978. *Why Not Say It Clearly? A Guide to Scientific Writing*. Boston: Little, Brown and Co.

Measurements and Statistics

Baily, N.T.J. 1981. *Statistical Methods in Biology*. 2nd edition. San Francisco: Freeman and Co. A good book for beginners.

Budiansky, Stephen, Art Levine, Ted Gest, Alvin P. Sanoff, and Robert J. Shapiro. 1988. "The Numbers Racket: How Polls and Statistics Lie." *U.S. News & World Report* 11 July 1988.

Lippert, H., and H.P. Lehmann. 1978. *SI Units in Medicine: An Introduction to the International System of Units with Conversion Tables and Normal Ranges*. Baltimore: Urban & Schwarzenberg. An extensive discussion of SI units and their use.

Moore, Randy. 1989. "Inching Toward the Metric System." *The American Biology Teacher* 51: 213–218.

Biological Literature

Ambrose, H.W., and K.P. Ambrose. 1987. *A Handbook of Biological Investigation*. Winston-Salem, NC: Hunter Textbooks, Inc.

Kronick, David A. 1985. *The Literature of the Life Sciences*. Philadelphia: ISI Press.

Morton, L.T., S. Godbolt (eds.). 1984. *Information Sources in the Medical Sciences*. 3rd edition. London: Butterworths.

Smith, R.C., W.M. Reid, and A.E. Luchsinger. 1980. *Smith's Guide to the Literature of the Life Sciences*. 9th edition. Minneapolis: Burgess Publishing Co. A detailed guide to biological literature.

Doublespeak

Fahey, Tom. 1990. *The Joys of Jargon*. New York: Barron's Educational Series.

Lutz, William. 1989. *Doublespeak*. New York: Harper Perennial.

Technical Guides for Biologists

American Medical Association, Scientific Publications Division. 1976. *Style Book: Editorial Manual*. 6th edition. Acton, MA: Publishing Sciences Group, Inc. Detailed instructions for preparing papers for journals published by the American Medical Association. Some recommendations differ from those in the CBE Style Manual.

CBE Style Manual Committee. 1978. *Council of Biology Editors Style Manual: A Guide for Authors, Editors, and Publishers in the Biological Sciences*. 5th edition. Washington, DC: Council of Biology Editors. This is a standard reference for all biologists.

A P P E N D I X   T W O

Words Frequently Misused by Biologists and Other Scientists

You learned in Chapter 3 the importance of always choosing the most precise word to convey meaning. Using words carelessly not only distracts readers from the author's message but also discredits the writer. Although poor usage and misspellings can be corrected by an editor or printer before the manuscript appears in a journal, it is the author's job to choose words carefully and accurately to best communicate with readers.

The words listed below are frequently misused. More information about word usage can be found in the reference books listed in Appendix One.

ability
capacity
Ability is the state of being able or the power to do something. *Capacity* is the power to receive or contain.

about
approximately
About indicates a guess or rough estimate (about half full). *Approximately* implies accuracy (approximately 42 m).

accuracy
precision
Accuracy refers to the degree of correctness, while *precision* refers to the degree of refinement. Thus, 5.845 is more precise than 5.8 but is not necessarily more accurate.

adapt
adopt
Adapt means *to change* or *adjust to. Adopt* means *to accept responsibility for, choose,* or *take from someone else.*

adequate
Adequate means enough. It is not synonymous with *plentiful* or *abundant.*

aggravate
The verb *aggravate* means *to intensify or increase.* Many writers incorrectly use *aggravate* also to mean to irritate or to annoy. People cannot be aggravated; conditions can be.

albumen
albumin
Albumen is the white of an egg. *Albumin* is any of a large class of simple proteins.

aliquot
sample
An *aliquot* is contained an exact number of times in something else. For example, 10 ml is an aliquot of 20 or 30 ml but not of 25 or 45 ml. A *sample* is a representative part of a whole.

alleviate
Alleviate means *to give temporary relief.* It implies that the underlying problem is still unresolved.

allude
refer
To *allude* to is to refer indirectly. To *refer* is to name.

alternate
alternative
Alternate means *every other one. Alternative* means *another choice; the second of two choices.*

altogether
all together
Altogether means *completely, entirely, whole. All together* means *as a group.*

among
between
Among applies to three or more persons or things; *between* applies to two.

amount
number
Amount refers to mass, bulk, or quantity. *Number* applies to a quantity that can be counted.

analogy
An *analogy* is a comparison between two objects or concepts. For example, rain is to snow as water is to ice.

and/or
Avoid this unsightly combination. If you mean *and, or,* or *both,* say so.

anthropomorphic
Ascribing human characteristics to nonhuman things.

anticipate
expect
Use *anticipate* when you mean looking forward to something with a foretaste of the pleasure or distress it promises. Use *expect* in the sense of certainty that it will occur. Anticipate a meeting; expect the sun to set.

anxious
eager
Anxious means *uneasy* or *apprehensive. Eager* is the word to describe earnestly wanting something.

as to whether
Just say *whether.*

as yet
Omit the first word.

bacteria
Plural of bacterium.

basically
An overused and generally unnecessary word.

because of
due to
Because of means *by reason of* or *on account of. Due to* means *attributable to.*

believe
think
Believe means *to actively accept as true. Think* means *to apply mental activity and power in considering a question without necessarily being sure of the answer.*

beside
besides
Beside means *next to. Besides* means *in addition to.*

bisect
dissect
Bisect means *to cut into two nearly equal parts.* To *dissect* is *to cut into many parts.*

can
may
Can means *to know how to* or *to be able to. May* means *to have permission to* or *to be likely to.*

carry out
perform
Carry out means *to remove from the room. Perform* means *to do.* Biologists do not *carry out* experiments; they *do* experiments.

center around
Impossible. You can only center on.

circadian
diurnal
Circadian is an adjective meaning *about 24 hours*. *Diurnal* is an adjective meaning *repeated or recurring every 24 hours*. *Diurnal* also means *chiefly active in daylight hours*.

compare to
compare with
Use *to* when comparing two unlike things. Use *with* when comparing two similar things.

connote
denote
Connote implies a meaning beyond the usual, specific meaning. *Denote* indicates the presence or existence of something.

constant
continual
continuous
Constant means *steady and unceasing*. Continual means *recurring frequently*. *Continuous* means *without interruption*.

criteria
Criteria is the plural form of the word *criterion*.

dalton
molecular weight
A *dalton* is a unit of mass equal to one-twelfth the mass of an atom of carbon-12. The *molecular weight* is a ratio of the weight of one molecule of a substance to one-twelfth the mass of an atom of carbon-12.

data
Data is the plural of *datum*. Data *are* important.

decimate
Decimate is from the Latin *decem*, meaning *ten*. *Decimate* means *to kill one in ten*, as the Romans did to control mutinous troops.

definite
definitive
Definite means *precise* or *clear*. *Definitive* means *final* or *complete*. Definite results are exact and unquestionable; definitive results will never be surpassed.

deductive reasoning
inductive reasoning
Deductive reasoning starts with the observation of general principles and uses these principles to explain particular details. *Inductive reasoning* uses details to make generalizations. What Conan Doyle often calls deductive reasoning in the Sherlock Holmes stories is actually inductive reasoning.

dilemma
A *dilemma* is a situation in which one faces two undesirable alternatives (note the prefix *di-*, implying two, as in *di*alogue).

disinterested
uninterested
Disinterested means *impartial; not influenced by personal interest or self-interest.*
Uninterested means *not interested.* Effective biologists are disinterested in their work.

distinct
distinctive
Distinct means *individual. Distinctive* means *special* or *unique.*

dose
dosage
A *dose* is the amount of a substance to be given at one time. *Dosage* is the regular administration of a substance in some definite amount.

ecology
environment
Ecology (often misused as a synonym for *environment*) is the study of the relationship between organisms and their environment. *Environment* refers to our surroundings.

effect
affect
Effect, when used as a noun, is *the result of an action.* When used as a verb, *effect* means *to cause* or *to bring about. Affect* is a verb meaning *to influence* or *to cause a change or an effect.*

enable
permit
Enable means *to make possible. Permit* means to *give consent.*

end product
Just say *product.*

enhance
increase
Enhance means to increase the contrast. *Increase* means to enlarge the value.

equally as
Equally as is redundant and always incorrect. Delete *equally* or *as.*

Erlenmeyer flask
There is no *h* in this word. The flask is named after Emil Erlenmeyer (1825–1909), a German chemist.

essential
Essential means *necessary for the existence of something else.* It is not synonymous with *important* or *desirable.*

except
accept
As a preposition, *except* means *other than. Accept* means *to receive* or *to agree with.*

fortunate
fortuitous
Fortunate means *having good luck. Fortuitous* means *happening by chance.*

farther
further
Farther refers to distance in space. *Further* applies to additional or advanced degrees or quantities.

fewer
less
Fewer applies to quantities that can be counted. *Less* applies to quantities that must be measured rather than counted. *Fewer* should be used only when an actual count can be made.

genus
Genus is the singular form of *genera*.

homogenous
homogeneous
Use *homogenous* as the adjectival form of *homogeny*—in biology, correspondence of organs or parts descending from a common origin. *Homogeneous* means *uniform throughout*.

incapable
Just say *unable*.

infer
imply
Infer means *to conclude or deduce from an observation or from facts*. *Imply* means *to express indirectly*. A speaker implies, and a listener infers something from what the speaker says.

irregardless
Use *regardless* or *irrespective*, not this bastard mixture of the two. The *ir-* is redundant; it means the same thing as the *-less* on the end of the word. Saying *irregardless* is like saying *irreckless*.

Krebs cycle
Often written incorrectly as *Kreb's cycle*. The cycle, also called the tricarboxylic acid cycle, is named after Sir Hans Adolph Krebs, its discoverer.

limit
delimit
Limit means *to restrict by fixing limits*. *Delimit* means *to determine or fix limits*.

locus
Locus means *the position of a gene on a chromosome* or *a set of points (loci) satisfying an equation*. In more general contexts, use *place, location,* or *position*.

media
Media is the plural of **medium**.

mitosis
cellular division
Mitosis refers to division of the nucleus. *Cellular division* is division of the cell.

probable
feasible
Probable means *likely to happen*. *Feasible* means *possible* but not necessarily *probable*.

principal
principle
Principal as a noun means *a sum of money* or *a chief person* and as an adjective, *chief, most important. Principle* means *a basic law.*

since
because
Use *since* to refer to time. Use *because* as the conjunction that introduces a reason.
  Since beginning the experiment I have . . .
  The data were duplicated because . . .

small
few
Small refers to the size of an object; *few* refers to the number of items of the same kind.

theory
hypothesis
A *theory* is a broad, integrated, and general concept supported strongly by scientific evidence and useful in predicting a wide range of phenomena. A *hypothesis* is a proposition for experimental or logical testing.

ultimate
penultimate
Ultimate means *last. Penultimate* means *next to last.*

unique
Unique means the only one of its kind. *Unique* cannot be modified and is not synonymous with *unusual, strange,* or *odd.*

APPENDIX THREE

The Library of Congress System for Classifying Books and Journals

QH 72	Nature Conservation. Landscape Protection		QK 475	Trees and Shrubs
			QK 494	Gymnosperms
QH 90	Water. Aquatic Biology		QK 495	Angiosperms
QH 91–95	Marine Biology		QK 520	Ferns
QH 96	Freshwater Biology		QK 534	Mosses
QH 201	Microscopy		QK 564–580	Algae
QH 301–349	General Biology		QK 581	Lichens
QH 351	Morphology		QK 600–638	Fungi
QH 425	Animals		QK 641–707	Plant Anatomy
QH 426–470	Genetics		QK 710–929	Plant Physiology
QH 471–499	Reproduction		QK 930–935	Physiographic Regions. Water.
QH 506	Molecular Biology			
QH 515	Photobiology		QK 936–939	Physiographic Regions. Land.
QH 540–559	Ecology			
QH 573–705	Cytology		QK 940–977	Physiographic Regions. Topographic Divisions.
QK 1–102	Botany			
QK 110–195	North American Botany		QL 1	Zoology

QL 362–599	Invertebrates		QM 601	Human Embryology
QL 605	Vertebrates		QP 1	Physiology
QL 514–638	Fishes		QP 88	Physiology of Tissue
QL 667	Amphibians		QP 99	Blood
QL 671	Birds		QP 101	Cardiovascular System
QL 700–739	Mammals		QP 141	Nutrition
QL 750	Animal Behavior		QP 186	Glands
QL 799	Morphology, Anatomy		QP 301	Movements
QL 951	Embryology		QP 351–495	Neurophysiology and Neuropsychology
QM 1	Human Anatomy			
QM 111–131	Skeleton		QP 501–801	Animal Biochemistry
QM 178	Vascular System		QP 901	Experimental Pharmacology
QM 301	Organ of Digestion			
QM 451	Nervous System		QR 1–74	Microbiology
QM 501	Sense Organs		QR 75–200	Bacteria
QM 531	Regional Anatomy		QR 201	Pathogenic Microorganisms
QM 550	Human and Comparative Histology		QR 355–484	Virology

A P P E N D I X F O U R

Typical Letter for Requesting Permission to Reproduce Material from Another Source

Date _____

To: Permissions Department

I am writing an article entitled _____
_____ that I will submit for
publication to _____. I would like your permission to
include the following information in my article:

_____
_____
_____

Volume _____ Page(s) _____ Year _____ from the article entitled

_____
_____

written by _____.

    If you grant me permission to use this material, I will credit the authors and
your journal as the source. I am sending a copy of this request to the (publisher or
author).

Sincerely,

_____
(Signature)

Permission Granted:

_____   _____
(Signature)   (Date)

A P P E N D I X   F I V E

Symbols and Abbreviations Used in Biology

Term/Unit of Measurement	Symbol or Abbreviation	Term/Unit of Measurement	Symbol or Abbreviation
calorie	cal	microliter	μl
centimeter	cm	micrometer	μm
complementary DNA	cDNA	milliliter	ml
		millimeter	mm
cubic centimeter	cm³	minute (time)	min
cubic meter	m³	molar (concentration)	M
cubic millimeter	mm³		
day	d	mole	mol
degree Celsius	°C	nanometer	nm
degree Fahrenheit	°F	number (sample size)	N
figure, figures	Fig., Figs.		
gram	g	parts per million	ppm
greater than	>	percent	%
hectare	ha	plus or minus	±
height	ht	probability	P
hour	h	ribosomal RNA	rRNA
joule	J	second (time)	s
kelvin	K	species (singular)	sp.
kilocalorie	kcal	species (plural)	spp.
kilogram	kg	square centimeter	cm²
kilometer	km	square meter	m²
less than	<	square millimeter	mm²
liter	l *or* L *or* liter *to avoid confusing with the numeral 1*	standard deviation	SD
		standard error	SE
logarithm (base 10)	log	standard temperature and pressure	STP
logarithm (base *e*)	ln		
mean	$\bar{x}$	transfer RNA	tRNA
messenger RNA	mRNA	volt	V
meter	m	watt	W
microgram	μg	weight	wt
		year	yr

APPENDIX SIX

Prefixes and Multiples of SI Units

Factor by Which the Unit is Multiplied	SI Prefix	Symbol
1 000 000 000 000 000 000 $= 10^{18}$	exa	E
1 000 000 000 000 000 $= 10^{15}$	peta	P
1 000 000 000 000 $= 10^{12}$	tera	T
1 000 000 000 $= 10^{9}$	giga	G
1 000 000 $= 10^{6}$	mega	M
1 000 $= 10^{3}$	kilo	k
100 $= 10^{2}$	hecto	h
10 $= 10^{1}$	deca	da
1 $= 10^{0}$ (unity)		
0.1 $= 10^{-1}$	deci	d
0.01 $= 10^{-2}$	centi	c
0.001 $= 10^{-3}$	milli	m
0.000 001 $= 10^{-6}$	micro	μ
0.000 000 001 $= 10^{-9}$	nano	n
0.000 000 000 001 $= 10^{-12}$	pico	p
0.000 000 000 000 001 $= 10^{-15}$	femto	f
0.000 000 000 000 000 001 $= 10^{-18}$	atto	a

A P P E N D I X   S E V E N

Authorization to Participate as a Subject in Research

It is the responsibility of the principal investigator to retain a copy of *each* signed consent form for at least five (5) years beyond the termination of the subject's participation in the proposed activity. Should the principal investigator leave the University, signed consent forms are to be transferred to the Institutional Review Board for the required retention period.

Project Title _____

Grant or Contract No. _____

Principal Investigator _____ Department

I consent to the performance upon _____
(myself or name of patient)

of the following treatment or procedure _____

_____

The purpose of the procedure or treatment: _____

_____

Possible alternative methods of treatment: _____

_____

Discomforts and risks reasonably to be expected: _____

_____

Benefits which may be expected: _____

_____

The nature and general purpose of the experimental procedure or treatment and the known risks have been explained to me and I understand them. I understand that any further inquiries I make concerning the procedure or treatment will be answered. I understand that my identity will not be revealed in any publication or document resulting from this research without my permission. I also understand that it is not possible to identify all potential risks in an experimental procedure; however, I believe that reasonable safeguards have been taken to minimize both the known and the potential unknown risks. This authorization is given with the understanding that I may terminate my service as a subject at any time after notifying the project director and without any prejudice. Reasonable and immediate medical attention, as exemplified by the services of the University Student Health Center, will be provided for physical injury caused directly by participating in this protocol. Any financial compensation for such physical injury will be at the option of the University, and decided on a case-by-case basis. In the event questions should arise concerning research-related injury to the subject or subject's rights, please contact the following for additional information.

_____ _____   _____
Principal Investigator or Contact Person   Phone   Signature (Subject's)

GLOSSARY

abstract   A summary of an article, book, or report.

active voice   Writing style in which the subject acts. Example: Charles Darwin wrote *The Origin of Species.* (See *passive voice*.)

adjective   A word that describes or limits the meaning of a noun or noun phrase. An adjective tells which, what kind of, or how many. Examples: *good* experiment, *red* reagent, *three* meters.

adverb   A word that modifies or expands the meaning of a verb, adjective, or other adverb. An adverb tells how many, when, or where. Adverbs often, but not always, end in *-ly*. Examples: work *slowly*, move *fast*.

antecedent   The word, phrase, or clause referred to by a pronoun. Examples: I read Gould's *essay* and liked it. (*Essay* is the antecedent of *it*.) A pronoun must agree with its antecedent in person, number, and gender. Thus, *it* is singular because *essay* is singular.

cliché   An expression that was once fresh but has become dull and stereotyped by overuse. Clichés make people laugh at you, not with you. Example: Avoid clichés *like the plague.*

dangling modifier   A modifier that cannot logically modify any word in a sentence. Example: *Having left in a hurry,* his experiment remained unfinished.

galley proofs   A preliminary reproduction of text made for checking spelling, spacing, format, and related items.

halftone   A printed photo having a range of tones.

independent clause   A clause that expresses a complete thought and thus can stand alone. Although sentences are independent clauses, most independent clauses are parts of sentences. For example, "He studied elephants in Africa, and he earned money by writing for biology magazines" consists of two independent clauses.

infinitive   The basic form of a verb, usually preceded by *to*. Examples: *to work, to study, to write.*

metaphor   a figure of speech containing an implied comparison. Example:

Our minds have adjusted to [the nuclear shadow under which we live], as after a time our eyes adjust to the dark. —Jimmy Carter

noun   A word that names things.

passive voice   Writing style in which the subject receives the action. Passive voice usually involves *to be* verbs such as *is, was,* and *were*. Example: The report *was written* by the biology student. (See *active voice*).

pronoun   A word that replaces a noun. Common pronouns include *he, she, they, we, them, I,* and *me.*

redundancy   An unnecessary repetition of meaning. Examples: *advance planning, active participation, present time.*

running head   A title repeated at the top of each page of a book or paper.

subject   The part of a sentence or clause that performs or, in the passive voice, receives the action of the verb.

verb   A word that expresses action or being. Examples: *write, study, is, were.*

INDEX